Green Energy and Technology

Climate change, environmental impact and the limited natural resources urge scientific research and novel technical solutions. The monograph series Green Energy and Technology serves as a publishing platform for scientific and technological approaches to "green"—i.e. environmentally friendly and sustainable—technologies. While a focus lies on energy and power supply, it also covers "green" solutions in industrial engineering and engineering design. Green Energy and Technology addresses researchers, advanced students, technical consultants as well as decision makers in industries and politics. Hence, the level of presentation spans from instructional to highly technical.

Indexed in Scopus.

Indexed in Ei Compendex.

Virendra Kumar Vijay · Rimika Kapoor ·
Sharang Dev Sharma · Vandit Vijay ·
Rahul Jain

Compressed Biogas (CBG)

Process, Technology and Applications

Springer

Virendra Kumar Vijay
Centre for Rural Development
and Technology
Indian Institute of Technology Delhi
New Delhi, Delhi, India

Sharang Dev Sharma
Centre for Rural Development
and Technology
Indian Institute of Technology Delhi
New Delhi, Delhi, India

Rahul Jain
Former PhD Scholar, Centre for Rural
Development and Technology
Indian Institute of Technology Delhi
New Delhi, Delhi, India

Rimika Kapoor
Former PhD Scholar, Centre for Rural
Development and Technology
Indian Institute of Technology Delhi
New Delhi, Delhi, India

Vandit Vijay
Former PhD Scholar, Centre for Rural
Development and Technology
Indian Institute of Technology Delhi
New Delhi, Delhi, India

ISSN 1865-3529 ISSN 1865-3537 (electronic)
Green Energy and Technology
ISBN 978-981-96-5696-7 ISBN 978-981-96-5697-4 (eBook)
https://doi.org/10.1007/978-981-96-5697-4

This Springer imprint is published by the registered company Springer Nature Singapore Pte Ltd.
The registered company address is: 152 Beach Road, #21-01/04 Gateway East, Singapore 189721, Singapore

Preface

The growing global demand for sustainable and clean energy solutions has brought renewable energy technologies to the forefront of development. Compressed biogas (CBG) also known as Biomethane/ Bio-CNG in different parts of the world, has gained significant attention for its dual benefits of waste management and clean energy production. For a country like India, with its abundant biomass and cattle resources, CBG offers an unparalleled opportunity to address energy security while reducing greenhouse gas emissions. Aligned with India's environmental commitments, such as achieving net-zero emissions by 2070 and advancing toward a circular economy, CBG presents a pathway to a fossil fuel-free and sustainable energy future.

The book opens by establishing a foundational understanding of biogas technology, introducing its core principles, the variety of feedstock sources, and the biological and chemical processes that drive biogas production. This sets the stage for a deeper exploration of the scientific and technical aspects that underpin the field. Following this, attention shifts to biogas quality standards and its wide-ranging applications. Emphasis is placed on the necessity of meeting stringent quality benchmarks to ensure the gas can be effectively and safely utilized across various sectors.

A substantial portion of the book is dedicated to the technologies involved in cleaning and upgrading biogas to meet compressed biogas (CBG) specifications. The discussion begins with conventional methods for removing impurities and progresses into more advanced purification techniques. These include pressure swing adsorption (PSA) and water scrubbing—two widely adopted technologies known for their effectiveness in enhancing biogas purity. Additional focus is given to membrane and cryogenic separation technologies, both of which offer unique advantages and growing industrial relevance in the CBG production landscape.

The narrative then moves into the domain of emerging and next-generation technologies. Innovation in biogas processing is examined through the lens of research and development, with particular attention to new methodologies that promise to increase efficiency, scalability, and cost-effectiveness. This forward-looking perspective continues with an exploration of trends and ongoing research shaping the future of CBG. These insights aim to inform not only current practices but also the long-term evolution of the sector.

The book concludes with a comprehensive overview of India's CBG journey. It highlights the nation's achievements in establishing CBG plants, the role of government policies and financial incentives, and the practical challenges encountered along the way. This closing section also presents a vision for India's emergence as a global leader in compressed biogas, aligned with the national mission of *Atmanirbhar Bharat* and the aspiration to become a developed nation by 2047, as envisioned by the Hon'ble Prime Minister.

This book aims to serve as a valuable resource for students, researchers, policymakers, entrepreneurs, and industry professionals seeking to understand and contribute to the growth of CBG. By exploring established and emerging technologies, it aspires to inspire innovation and collaboration, paving the way for a greener and more sustainable energy future for India and the world. However, in many ways, the book is a collective outcome of the knowledge, inspiration, and encouragement received from several remarkable individuals—mentors, pioneers, colleagues, and collaborators—whose contributions to the field of compressed biogas (CBG) technology have shaped both the vision and content of this work.

The profound legacy of **Dr. K. C. Khandelwal**, former Adviser to MNRE and one of India's earliest champions of biogas technology, remains an enduring source of inspiration. His pioneering work continues to guide the direction of research and policy even today. Fond remembrance and deep respect are owed to **Prof. P. L. Dhar and the Late Prof. Rajendra Prasad** and **the Late Prof. R. R. Gaur of IIT Delhi**, and **the Late Prof. A. N. Mathur**, CTAE, Udaipur, whose academic foresight and unwavering belief in sustainable energy solutions have left an indelible mark on the field and on all who had the privilege to work with them.

The Centre for Rural Development and Technology (CRDT), IIT Delhi, has long been a nurturing ground for ideas and action. In this regard, heartfelt appreciation is extended to **Prof. Vivek Kumar**, **Prof. Priyanka Kaushal**, **Prof. Pooja Ghosh**, and **Late Prof. Ram Chandra** for their scholarly insight and constant support over the course of writing this book. Special thanks are also due to **Prof. P. M. V. Subbarao** of the Mechanical Engineering Department, IIT Delhi, and **Prof. Deepak Sharma**, CTAE, MPUAT, Udaipur, for their technical rigor and deeply insightful contributions to the understanding of biogas enrichment and bottling technology developed by IIT Delhi.

The practical and industry-oriented perspectives of **Mr. Sanjeev Nagpal**, Managing Director, Sampurn Agri Ventures Pvt. Ltd., **Mr. Ashish Kumar**, Managing Director, VERBIO India, **Mr. Sanjay Ganjoo**, Director General, Indian Federation of Green Energy, and **Mr. Y. B. Ramakrishna**, Sr. Vice President, IFGE and Former Chairman Working Group on Biofuel, Ministry of Petroleum and Natural Gas, GoI have been vital in bridging the gap between research and real-world application. Their inputs continue to enrich this evolving sector. A special note of gratitude goes to **Shri Kanwar Singh**, General Manager, and **Shri Mahesh Girdhar**, Managing Director and CEO of EverEnviro Resource Management Private Limited, for their leadership and continued commitment to fostering innovation and implementation in the CBG sector.

Recognition is also extended to **Shri G. L. Meena**, Former Advisor, **Sangita M. Kasture**, Scientist G, **Dr. Gaurav Mishra,** Scientist F and **Shri S. R. Meena**, Scientist E, at MNRE, for their valuable policy guidance. We extend our heartfelt thanks to **Shri Pradip Kumar Das**, CMD, Indian Renewable Energy Development Agency, for his insights regarding the financial ecosystem of the CBG sector in India and longstanding support in the national efforts to scale up biogas initiatives. Grateful acknowledgment is made to **Mr. Amit Agrawal**, former BDTC Program Manager, IIT Delhi, for their strategic inputs and active involvement in capacity building and outreach. And finally, a sincere and heartfelt appreciation for the **Ph.D. research scholars of the Biogas Laboratory, CRDT, IIT Delhi**. Their tireless pursuit of knowledge, experiments, and innovation forms the very bedrock of much that this book seeks to share.

This work owes gratitude to each of these individuals—not only for their contributions to the field but also for their enduring belief in the potential of biogas to transform India's energy landscape. Lastly, we sincerely thank the **family members of all authors** for their unwavering support and encouragement. A sincere note of gratitude to **Mrs. Pramila Vijay**, wife of **Prof. V. K. Vijay**, for her constant inspiration and understanding.

New Delhi, India

Virendra Kumar Vijay
Rimika Kapoor
Sharang Dev Sharma
Vandit Vijay
Rahul Jain

Competing Interests The authors have no competing interests to declare that are relevant to the content of this manuscript.

Contents

About the Authors

Prof. Virendra Kumar Vijay is a professor at the Centre for Rural Development and Technology, Indian Institute of Technology (IIT) Delhi, and is an acclaimed scientist in rural technologies, transforming the higher education process, biogas enrichment and bottling technology for vehicular application (Compressed Biogas-CBG), bioenergy, etc. He has more than three decades of Teaching, Research, and Outreach experience in renewable energy, rural development and technologies, biogas generation, enrichment and bottling technology for vehicular application, cow-based economy, etc. Prof. Vijay is the national coordinator of Unnat Bharat Abhiyan—a flagship program of the Ministry of Education, Government of India, and coordinator of the Biogas Development and Training Centre (BDTC), Ministry of New and Renewable Energy, Government of India. He has been known for designing and developing small-scale biogas purification CBG and bottling technology for vehicular applications in India since 2002. IIT Delhi patented the water scrubbing technology in 2006 and licensed it to more than 11 industries, many Gaushalas, and young entrepreneurs for field implementation. He has also been the General Secretary of Biogas Forum India. Prof. Vijay has over 240 publications in journals and conferences, several books, and a lab manual to his credit. He has also delivered over 250 invited lectures and talks and received numerous awards during his exemplary academic career.

Dr. Rimika Kapoor holds a Ph.D. and a postdoctoral fellowship from the Indian Institute of Technology (IIT) Delhi. She acquired her master's in Engineering Innovation and Entrepreneurship from Toronto Metropolitan University, Canada, and an MBA from IMT Ghaziabad. She has over 20 publications in international journals and conferences. Dr. Kapoor has received several awards, including the Esch Award (Canada) on bioplastics-based organic sanitary pads and the 7th ENERTIA Award. She is the founder of Green Loop Clean Tech and is the principal innovator and waste management expert. As the founder of GreenLoop CleanTech, she continues to lead advancements in carbon accounting, ESG, life cycle assessment, and clean technologies, working with businesses and academic institutions to promote sustainable solutions globally.

Mr. Sharang Dev Sharma is a Ph.D. research scholar at the Centre for Rural Development and Technology (CRDT) at the Indian Institute of Technology (IIT) Delhi. His doctoral research currently focuses on optimizing the site suitability of compressed biogas (CBG) plants using Geographic Information Systems (GIS) and multi-criteria analysis tools. Mr. Sharma completed his Bachelor of Engineering in Electrical and Electronics Engineering. He further advanced his expertise by earning a Master of Engineering in Information and Communication Technologies from the Asian Institute of Technology, Bangkok, Thailand. Before starting his doctoral studies, Sharang worked as a Planning Manager for the Seagate Technologies External HDD project for Asia at Cal-Comp Electronics Public Company Thailand Limited, part of the New Kinpo Group in Taiwan.

Dr. Vandit Vijay is working as a *"Scientist C"* at the Sardar Swaran Singh National Institute of Bioenergy, Kapurthala (SSS-NIBE, an autonomous institute of the Ministry of New and Renewable Energy, Government of India). He worked as an outstation postdoctoral research scientist with TU Delft, Netherlands, from 2019 to 2021 in their Negative Emissions Program. He completed his Ph.D. from IIT Delhi from 2014 to 2019 on self-reliant energy systems for villages using locally available resources—ग्राम ऊर्जा स्वराज model. His major work areas at SSS-NIBE include biomass resource assessment and mapping, development of the National Biomass Atlas of India, biomass supply chains, and policies. His research also focuses on biomass utilisation through different conversion pathways for energy sufficiency in rural areas and carbon neutrality. He has experience in biogas and biodiesel production from different feedstocks and their utilisation for various energy applications, including biofuel-based IC Engines and sustainability analysis of bioenergy systems. He has more than 20 authored/co-authored publications. He has a B.Tech. degree in Mechanical Engineering (2008–2012), and before joining the Ph.D. program, he worked at the Society of Indian Automobile Manufacturers from 2013 to 2014.

Dr. Rahul Jain is an alumnus of the Indian Institute of Technology (IIT), Delhi. He holds an industrial Ph.D. in Environmental Engineering from IIT Delhi, conducted in collaboration with the National Thermal Power Corporation (NTPC) Ltd., under the prestigious Prime Minister's Doctoral Research Fellowship. He also earned an MS in Energy from IIT Guwahati. Dr. Jain has made notable contributions to environmental science and technology. His research, published in several journals and presented at many international forums, contributes to bridging the gap between technology and policy, particularly in waste management, circular bioeconomy, bioremediation, climate change mitigation, and carbon capture. Dr. Jain was honoured with the International Biogas Industry Awards 2024 by WBA.

Chapter 1
Fundamentals of Biogas Technology

1.1 Historical Insight on Anaerobic Digestion Technology

The discovery of anaerobic digestion dates back to the seventeenth century when Plinius discovered lights under the swamps. It was no longer than 1630 when Van Helmont found flammable gases evolving from decaying organic matter [1]. In 1776, Volta learned that marsh gas formed an explosive mixture with air, and its quantity was governed by the amount of decaying material [25]. However, a breakthrough was established in 1804 by Dalton, who determined methane's chemical composition. Henry, in 1806, concluded that methane was much like the marsh gas that Volta defined [7]. Later, in 1808, Sir Humphry Davy discovered methane from cattle manure. These early discoveries led to the industrialization of anaerobic digestion. Later, in 1856, Reiset observed methane liberation from decaying piles of manure and recommended studying and thoroughly understanding the decomposition process of organic materials.

The first biogas plant based on sewage waste in India was installed in 1859 at a leper colony in Bombay. As a result of scientific interest in the technology at the end of the 1900s, it was discovered that microbial activity and methanogenesis were related. However, in late 1876, it was discovered that acetate can be converted to CH_4 and CO_2 stoichiometrically. Finally, in 1884, Gayon, a French student, generated 100 L of methane per m^3 of manure by fermenting it at 35 °C [34]. In 1895, the UK lit the streetlamps using the biogas obtained from a sewage sludge-based biogas plant in Exeter. In 1897, another biogas plant based on human waste was built in Leper Asylum in Matunga, Bombay, where the biogas was utilized for lighting [10]. In the 1900s, biogas production setups were established in the UK and Germany to handle sewage. Many centralized drainage systems were set up throughout Europe to use waste to produce biogas. At that point, anaerobic digestion was being considered as a way to lower the overall amount of waste. In the early twentieth century, the Imhoff tank, an early form of the digester, was patented in Germany.

V. K. Vijay et al., *Compressed Biogas (CBG)*, Green Energy and Technology,
https://doi.org/10.1007/978-981-96-5697-4_1

Awareness of anaerobic digestion increased during the 1930s, as microbiology became recognized. It drove the research to understand the parameters needed to optimize AD. In early 1936, researchers briefed their observations on the fermentation of agricultural residues such as straw and silage. In 1937, biogas as compressed gas was tested as a vehicular fuel for the first time in a vehicle [29]. However, an increase in oil prices motivated research in the biogas sector. In India and China, biogas has been produced using animal manure and kitchen waste and has been utilized as fuel for cooking and lighting for a long time. Animal-based biogas plants were used in Bombay to produce biogas for lighting and cooking. In the 1960s, Khadi and Villages Industries Commission (KVIC) designed a floating dome (made of steel) type biogas plant to support an already running outreach program of the Government of India, to provide cooking fuel to villagers. China started a similar program in the 1960s, and by the early 1980s, the country had built around 5 million biogas plants. The biogas market is being intensely encouraged and has increased considerably recently. With significant governmental and public support, research and development in this sector have led to innovative and advanced biogas production processes and technologies for its effective utilization.

1.2 Microbial Physiology and Biochemical Processes in Anaerobic Digestion (Microbial Degradation Steps Leading to Methane)

Anaerobic digestion (AD) is a microbiologically facilitated process in which feedstock biopolymers convert organic carbon into methane and carbon dioxide without oxygen [6]. It is a complex biochemical method to treat organic wastes ranging from complex lignocellulosic materials to easily degradable food wastes anaerobically in a specialized designed tank or vessel called the *anaerobic digester*. The process is called *anaerobic digestion,* as shown in Fig. 1.1**.** This process produces renewable energy in biogas (because of the biological nature of gas production) and a nutrient-rich slurry in the form of *digestate.* It can be directly used as a fertilizer on farming land or processed and commercialized as organic fertilizer for improving soil health (Al Seadi et al. 2008).

The biogas production process is an outcome of interconnected processes (steps), in which the complex polymers as a feed are constantly broken down into smaller and simpler monomers. Particular sets of microorganisms are responsible for the decomposition and breakdown of larger molecules at every step [41]. A basic illustration of the AD process is displayed in Fig. 1.2. It comprises four main processes: *hydrolysis, acidogenesis, acetogenesis, and methanogenesis* [36]. In anaerobic digestion, these activities co-occur in both space and time. The digester has a side opening (charge pit) for feeding organic waste feedstock and a cylindrically shaped container above to collect the gas.

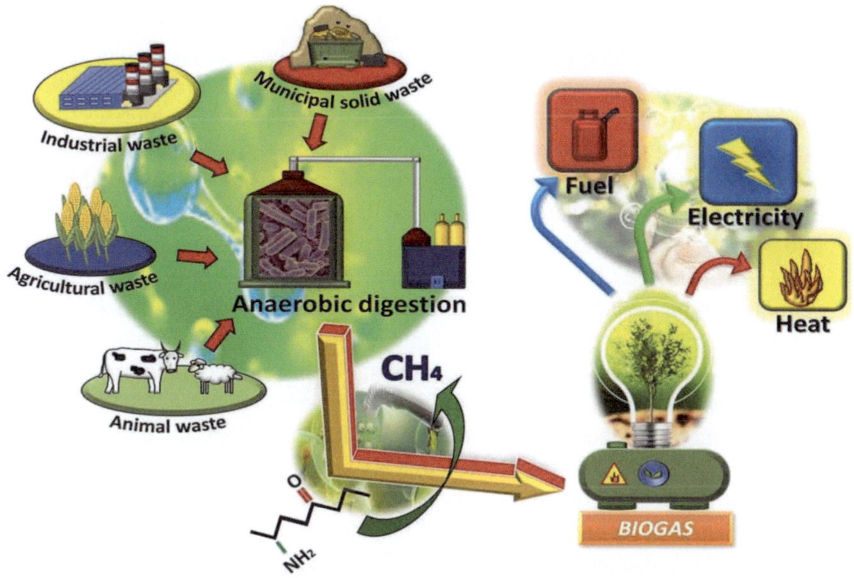

Fig. 1.1 Biogas production from different waste and its application. Reproduced under license CC-BY-4.0 [26]

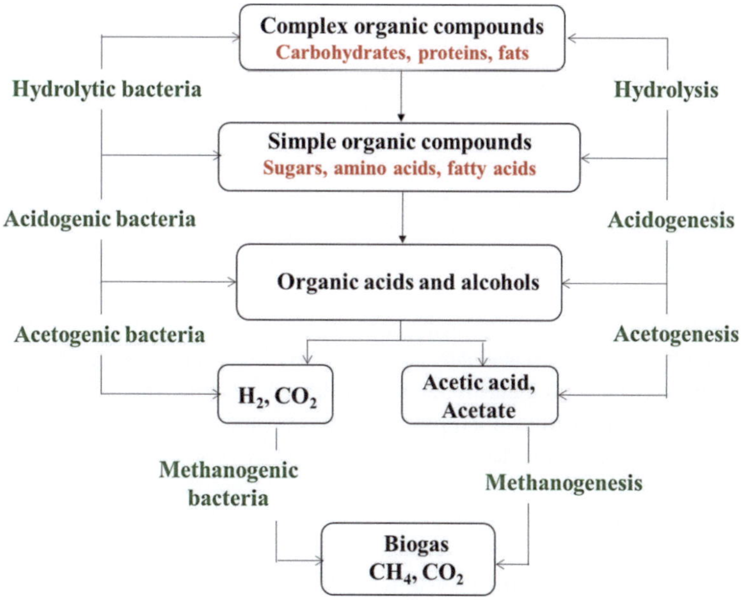

Fig. 1.2 Biochemical processes in anaerobic digestion. Reproduced with Permission from Elsevier [38, 37]

1. **Hydrolysis**: This is the first phase of the AD procedure. It makes complicated polymeric organic molecules easier to break down into simpler molecules like sugar, fatty, and amino acids. To create the slurry, the feedstock is first liquefied in water. The first class of microorganisms, primarily anaerobes, comprises various facultative or obligatory fermentative bacterial groups, sometimes referred to as hydrolytic bacteria. These bacteria are known as acid formers, and with the aid of extracellular enzymes, they hydrolyze (solubilize) complex chemical molecules into simpler, low molecular weight compounds like alcohols and short-chain fatty acids. At this stage of decomposition, the predominant hydrolytic enzymes are cellulase, protease, lipase, xylanase, and amylase. According to Adekunle and Okolie [3], protease converts proteins into amino acids, lipase breaks lipids into fatty acids, and polysaccharides into sugars. For instance, cellulolytic bacteria, which release the cellulose enzyme, first break down cellulose, which is made up of polymerized glucose units, into dimers (disaccharides) and then monomers (monosaccharides), such as glucose.
2. ***Acidogenesis***: It is regarded as AD's second stage. At this point, the digester's environment becomes acidic because of the activity of acidogenic (fermentative) bacteria, which also create shorter volatile fatty acids, ammonia (NH3), hydrogen (H_2), carbon dioxide (CO_2), hydrogen sulfide (H_2S) and organic acids, including lactic, propionic, butyric, and acetic acids. The ensuing products are still unsuitable for producing methane since they are large and complex. The second type of bacteria, which are obligatory hydrogen producers and facultative anaerobic bacteria, generates acidogenic bacteria that change the monomers (amino acids, fatty acids, and sugar) into carbon dioxide, hydrogen, and acetogenic dehydrogenation. The microorganisms involved in acidification are *Psedomonas formacans, Bacillus cereus, Bacillus megaterium* and *Clostridium carnofeetidium*, etc. [14]. The metabolic pathway of the step is shown in Table 1.1.
3. ***Acetogenesis***: In the acetogenesis state, the products of the acidogenesis stage are transformed into methanogenic substrates. Acetogenic bacteria which contribute to acetogenesis, include *Syntrophomonas wolfei, Syntrophobacter wolinii, Clostridium* spp., *Peptococcus anaerobes, Actinomyces,* and *Lactobacillus* [35] (Table 1.1). During this stage, acetogens oxidize alcohols and volatile fatty acids into acetates, acetic acid, hydrogen gas, and carbon dioxide.
4. ***Methanogenesis***: Methane is created from the intermediate products derived from the anaerobic digestion process at this last and most important stage, with the end products of the acetogenesis stage, hydrolysis, and acidogenesis stage. Microorganisms like methanogens are active during this stage, which is very sensitive to pH changes. Methanogens are a distinctive group of bacteria and play an important role in converting substrate into methane. They are obligate anaerobes and have a slow growth rate [6]. Effective degradation is impossible without methanogens due to the accumulation of fatty acids and alcohol products of the fermentative and other hydrogen-utilizing bacteria, which have almost the same energy content as the original organic matter. Main methanogens in anaerobic digestion fall into two main groups: acetoclastic-acetate degrading (*Methanosarcina barkeri, Metanonococcus mazei, Methanothrix soehngenii*)

Table 1.1 Metabolic pathway of steps in anaerobic digestion [14, 35, 39]

Process	Metabolic Pathway	Bacteria Dominant
Step I: Hydrolysis	• Proteins to soluble peptides and amino acids • Carbohydrates to soluble sugars • Lipids to fatty acids or Alcohols $(C_6H_{10}O_5)_n + nH_2O \rightarrow n(C_6H_{12}O_6) + H_2$	*Clostridium, Proteus vulgaris, Vibrio, Bacillus, Peptococcus, Bacteriodes, Acetovibrio, celluliticus, Staphylococcus, Bacteriodes*
Step II: Acidogenesis	• Amino acids to fatty acids, acetate, and NH_3 • Sugars to intermediary fermentation products $C_6H_{12}O_6 + 2H_2O \rightarrow 2CH_3COOH + 2CO_2 + 4H_2$ $3C_6H_{12}O_6 \rightarrow 4CH_3CH_2COOH + 2CH_3COOH + 2CO_2 + 2H_2O$ $C_6H_{12}O_6 \rightarrow CH_3CH_2CH_2COOH + 2CO_2 + 2H_2$ $C_6H_{12}O_6 \rightarrow 2CH_3CHOHCOOH$ $C_6H_{12}O_6 \rightarrow 2CH_3CH_2OH + 2CO_2$	*Lactobacillus, Escherichia, Bacillus, Staphylococcus, Pseudomonas, Sarcina, Desulfovibrio, Selenomonas, Streptococcus, Veillonella, Desulfobacter, Desulfuromonas Clostridium, Eubacterium, limosum, Streptococcus*
Step III: Acetogenesis	• Higher fatty acids or alcohols to hydrogen and acetate • Volatile fatty acids and alcohols to acetate or hydrogen $CH_3CH_2COOH + 4H_2O \rightarrow CH_3COOH + {}_6H_2 + {}_2CO_2$ $CH_3CH_2COOH + 2H_2O \rightarrow CH_3COOH + {}_3H_2 + CO_2$ $CH_3CH_2CH_2OH + {}_3H_2O \rightarrow CH_3COOH + CO_2 + {}_5H_2$	*Clostridium, Syntrophomonas wolfei, Syntrophomonas wolinii*
Step IV: Methanogenesis	• Acetate to methane and carbon dioxide • Hydrogen and carbon dioxide to methane $2CH_3CH_2OH + CO_2 \rightarrow 2CH_3COOH + CH_4$ $CH_3COOH \rightarrow CH_4 + CO_2$ $CO_2 + 4H_2 \rightarrow CH_4 + 2H_2O$ $CH_3OH \rightarrow CH_4 + H_2O$	Acetolactic-acetate degrading bacteria (*Methanosarcina barkeri, Metanonococcus mazei, Methanothrix soehngenii*) Hydrogenotrophic-hydrogen utilizing bacteria (*Methanobacterium, Methanoculleus, Methanospirillum*)

and hydrogenotrophic-hydrogen utilizing (*Methanobacterium, Methanoculleus, Methanospirillum)* [39] (Table 1.1). Acetate is one of the substrates of the acetoclastic pathway, and hydrogen with carbon dioxide is a general substrate of the hydrogentrophic pathway [35]. At this point, acetate accounts for 70% of the methane produced, with the remaining 30% coming from the conversion of carbon dioxide (CO_2) and hydrogen (H_2). In this last stage of anaerobic digestion, methane-producing bacteria (methanogens) convert acetate, H_2, and CO_2 into methane (CH_4), carbon dioxide (CO_2), water, and other products. It comprises the slowest biochemical reactions, which are highly sensitive to various factors such as the composition of the feedstock, rate of feeding, temperature, pH, etc.

The slowest reaction of the anaerobic process assesses the biogas production rate. The hydrolysis step, which produces toxic by-products (complex heterocyclic compounds) or undesirable volatile fatty acids (VFAs), is the rate-limiting step for complex organic substrates containing cellulose, hemicellulose, and lignin. At the same time, methanogenesis is the rate-limiting step for readily biodegradable substrates, such as wastewater with a high fraction of soluble organic compounds [3]. A fast rate of hydrolysis results in the partial decomposition of polymeric compounds, which eventually reduces pH below 7.0 due to an increase in the concentration of the VFAs. This acidic condition is inhibitive for the methanogenic bacteria. The hydrolytic stage limits methane production if the acidogenic phase is too fast. Complex compounds like cellulose, fats, or proteins decompose in several days into monomers, unlike simple carbohydrates, which hydrolyze within a few hours. Hence, the process must be designed according to the characteristics of the substrates to achieve complete degradation without process failure.

1.3 AD Operating Parameters

Several critical factors affect the efficiency of the anaerobic digestion (AD) process. Different feedstock decomposes at different rates and produces different amounts of biogas depending upon the methane potential, biological degradability, carbon and nutrients availability, and moisture content. Therefore, it is necessary to provide optimum conditions for microorganisms to achieve efficient AD. AD processes are susceptible to parameters such as anaerobic environment, temperature, pH, stirring, retention time, etc., as they considerably affect the growth and activity of the microorganisms. Even though the deciding elements interact, there is a lack of accurate quantitative data on gas output as a function of the aforementioned parameters. The qualitative effects of these factors on anaerobic digestion are discussed below.

A. **Temperature**

Temperature is a major parameter affecting the anaerobic activity of microorganisms. For an efficient digestion process, constant temperature is one of the main requirements as the deviation in temperature negatively affects biogas production. Microorganisms display the best growth and metabolic rates in a distinct temperature range which is different for different species. Microorganisms are grouped as psychrophilic, mesophilic, and thermophilic depending upon the temperature range in which they are active. The optimum temperatures for psychrophilic, mesophilic, and thermophilic microorganisms are 10 °C, 20–45 °C, and > 50 °C, respectively [9]. Low psychrophilic temperatures lead to a reduction in slow degradation (70–80 days minimum retention time), substrate utilization rate, VFA production, ammonia concentration, and metabolic rate of microorganisms. This results in an increase in the time required for digester start-up and a decrease in biogas yields [22]. Bacteria show stability and relatively high strength in the mesophilic temperature range. The minimum retention time during this temperature range is 30–40 days. This results in appropriate decomposition of the substrate and high methane yields. Mesophilic processes are tolerant to temperature fluctuations in the range of $\pm$ 3°.

During thermophilic conditions, the methanogenic diversity is low, making the process less stable and energy demanding. Thermophilic anaerobes are more sensitive to fluctuations in temperature, so they need a longer time to adapt to a new temperature. The growth rate of methanogenic bacteria is high at thermophilic conditions, resulting in a faster and more efficient process with a short retention time of 15–20 days. Consequently, a thermophilic digester can be either heavily loaded or run at a lower hydraulic retention time (HRT) than at mesophilic conditions. In addition, the production rate of methane in the thermophilic AD process is approximately twice that in the mesophilic AD process, so the requirement of the volume of the digester in thermophilic conditions is reduced to half of the volume of the digester used in the mesophilic condition. Warm (> 55 °C) soluble wastes give eight times higher volumetric gas yields per volume of digester/day. However, thermophilic digestion is not cost-efficient, as it consumes energy to raise the reactor's temperature to 55 °C.

The major disadvantage of the thermophilic process is ammonia inhibition. At high temperatures, ammonia toxicity increases, which washes out the microbial population [24]. An increase in the concentration of volatile fatty acids (VFAs) leads to a drop in pH, which partly counteracts the effect of ammonia. However, very high temperatures decrease the metabolic rate due to the degradation of enzymes that are very essential for the life of the bacterial cell. The availability of hydrogen also negatively affects the metabolism of acetogenic bacteria. So, it is desired to keep the partial pressure of hydrogen low for efficient working of the acetogenic and H_2-producing bacteria. The advantages of thermophilic and mesophilic AD can be incorporated into two phases anaerobic digestion process; thermophilic hydrolysis/acidogenesis and mesophilic methanogenesis may provide the optimal conditions for the anaerobic digestion process.

B. **pH**

In anaerobic digestion (AD), pH control is necessary for the activity of the methanogens. The ideal pH level of an anaerobic digester should be in the range of 6.7–7.5; however, methanogens work best in a pH between 7 and 7.5. pH varies throughout the process due to the production of the various product(s) in various stages, i.e., the production of volatile fatty acids (VFAs) during the acidogenesis stage reduces the pH. This low pH inhibits the activity of methanogenic bacteria in the methanogenesis stage. The process is strictly inhibited if the pH is below 6.0 or rises above 8.5 [22]. The methanogenic microorganisms are susceptible to pH. Low pH halts microbial reactions during the digestion process. The accumulation of ammonia during protein degradation increases pH, while the accumulation of VFA decreases it. The high concentration of VFA can inhibit the activity of methanogens in the methanogenesis process [18]. However, the buffer capacity of the substrate compensates for the accumulation of VFA and eventually balances the pH, as in the case of animal (cattle and pig) manure, the surplus alkalinity of substrate balances the pH during VFA accumulation.

Hence, continuous monitoring and adjustments are required to fix a disturbed pH condition in an anaerobic digester. The foremost step is to stop feeding to slow down the activity of bacteria involved in fermentation and decrease acid production. This will increase the pH to an acceptable level of more than 6.8. Once the pH stabilizes, feeding can be started again at the low quantity and then increased slowly to avoid any further pH drop. The other technique is to raise the pH and provide more buffer capacity by adding chemicals like potassium, sodium hydroxide, sodium bicarbonate, sodium carbonate, ammonium hydroxide, lime, and sodium hydroxide [27]. After stabilization, the pH will be between 7 and 8.5.

C. **Seeding or Bacterial Population**

Acetogenic and methanogenic bacteria are inherently present in cattle with very low populations. Acetogenic bacteria multiply rapidly and increase their population, whereas methanogenic bacteria increase their population slowly. Therefore, anaerobic digestion of organic materials could be enhanced by using a small amount of slurry from a functioning digester as the inoculum for seeding. This slurry contains a good population of acetogenic and methanogenic bacteria. The seeding material can be mixed with this inoculum in a 1:1 to 1:3 or more ratio per requirement [5]. An increased quantity of inoculum above the permissible ratio results in low biogas production due to low input fed to the anaerobic digester [19].

D. **Retention Time**

Retention time is the time needed for organic materials to decompose completely. It is influenced by the kind of feedstock, the rate of organic loading, the temperature of the process, and the makeup of the substrate. It is connected to the rate at which microorganisms proliferate. Solids retention time (SRT) and hydraulic retention time (HRT) are the two types of retention times used in anaerobic digestion. The term "SRT" describes how long the solids (bacteria) stay inside the digester. Conversely, HRT is the amount of time the slurry needs from the moment of admission to the moment of exit following full deterioration inside the digester.

It is typically expressed in hours or days. Low HRT generally leads to VFA accumulation, resulting in bacterial washout and inhibition and even process failure, whereas a longer HRT results in inadequate utilization of feedstock and requirement of larger digesters [40]. HRT is calculated as the ratio of the volume of the digester (V_R) and the amount of substrate fed per unit time, as shown in the following equation:

$$\mathrm{HRT} = V_R / V$$

where

HRT hydraulic retention time [days]
V the volume of substrate fed per time unit [m^3/d].
V_R digester volume [m^3].

HRT typically lasts for 15–30 days. The HRT might be as high as 100 days in colder areas compared to 30–50 days in warmer climates [19]. Substrates like industrial wastewater containing simple biopolymers like sugar and starch-rich substrates do not require hydrolysis and require shorter retention times. While complex plant feedstock containing lignin, hemicellulose requires longer retention times. Hydrolysis regulates the breakdown process of these substrates and necessitates retention periods ranging from 50 to 100 days to guarantee adequate digestion and steady functioning.

E. **Organic Loading Rate**

Organic loading rate (OLR) is the amount of raw material (kg of volatile solids) fed into a digester per volume and per unit time as per the following equation.

$$\mathrm{OLR} = m * c / V_{\mathrm{R}}$$

OLR organic loading rate (kg/d*m^3).
c concentration of organic matter (%).
m mass of the substrate fed per unit time (kg/d).
V_{R} digester volume (m^3).

OLR is a significant parameter that affects the anaerobic digestion process, particularly under continuous feeding mode. It is a parameter to measure the biological conversion capability of the anaerobic digestion system. Between 0.5 and 2 kg of total volatile solids per unit volume of the digester per day is the ideal range for OLR [17]. Overloading the system above its OLR affects the digestion process. Feeding a large volume of fresh feed daily may temporarily inhibit bacterial activity during the early stages of the fermentation process. This leads to longer hydrolysis and acidogenesis than methanogenesis and thus results in a rise in VFA production, ultimately leading to irreversible acidification. It results in low biogas yields due to the accumulation of inhibitory acidic substances and undecomposed material such as fatty acid and eventually process failure. This reduces the pH, and the hydrolysis process is inhibited such that the methanogenesis bacteria are restricted and are not able to convert as much VFA to methane [40]. This creates an imbalance in the entire decomposition chain. The raw material type, retention duration, and process temperature make the selection. An excessively low oxygen yield ratio (OLR) during anaerobic digestion may lead to microorganism malnourishment, thereby decreasing the process's effectiveness. Methane production can be increased by starting at a low OLR and gradually increasing it to the optimal OLR. However, a low OLR and a long HRT are the best approaches to achieving a steady and elevated methane output. Understanding the substrates' dry solid and volatile compositions is necessary to calculate optimal OLR accurately.

F. **Mixing/Stirring**

Mixing the substrate in the digester is essential to maintain stability and homogeneity throughout the digester's volume. It facilitates the proper mixing of fresh incoming feed material with the already existing microorganisms in the digester and avoids thermal stratification, the formation of aggregates and scum. It promotes uniform temperature distribution and contact between microorganisms, substrates, and nutrients. Mixing prevents microorganisms from being washed out by the outflowing slurry. It lowers sedimentation, too, and hence decreases the risk of foaming. Mixing with the mechanical stirrers and recirculation of digested slurry with centrifugal pumps are the techniques that are commonly used for mixing in the digester.

G. **C/N Ratio**

The two most important components in the anaerobic digestion process are carbon and nitrogen. The organic feedstock primarily contains nitrogen in the form of proteins, ammonia nitrates, and carbon in the form of carbohydrates. The nutrient level, which is a crucial component of effective anaerobic digestion, is represented by the C/N ratio. Since carbon is used by bacteria 30 times more quickly than nitrogen, there should be 20–30 times as much carbon present as nitrogen for optimal nutrient availability. The ideal C/N ratio is thought to be between 20 and 30, and 25 C/N is frequently used in various types of studies [16]. Methanogens use nitrogen in the feedstock to meet their protein needs. The synthesis of enzymes, protoplasm, and amino acids is aided by nitrogen. It is transformed into ammonia, which, being a potent base, neutralizes the volatile acids that fermentative bacteria make. This aids in preserving the neutral pH conditions necessary for the process. In a high C/N ratio, methanogens consume nitrogen rapidly, eventually reducing biogas production. However, a low C/N ratio represents a high amount of nitrogen, and it produces toxic and inhibitory effects due to a high pH level of more than 8.5 due to excessive ammonia formation in the digester. It also lowers the biogas production. The C/N ratio of feedstock with a high C/N ratio can be optimized by co-digestion with a lower C/N ratio. The C/N ratio of different feedstocks is shown in Table 1.2.

H. **Ammonia Inhibition**

Anaerobic decomposition of nitrogen-containing organic substrates such as protein produces free ammonia (NH_3) and ammonium ions (NH_4^+). Ammonia plays a significant role in balancing the C/N ratio, which is one of the important parameters for the

Table 1.2 Typical C/N ratios of different feedstocks [16]

S. No.	Feedstock	*N* (%)	C/N ratio
1	Animal dung (cow, buffalo, sheep, pig, horse)	1.4–3.8	15–40
2	Poultry manure	6.3	5.2
3	Night soil	6	6–10
4	Slaughterhouse waste	7–14	2–4
5	Bagasse, wheat, and rice straw	0.3–0.5	120–150
6	Kitchen vegetable scraps	3.3	16
7	Food waste	2–4	2–18
8	Grass trimmings	2–4	12–16

AD process [41]. At optimum levels, it is a necessary nutrient for bacterial growth but can be toxic to the microbial community at high concentrations. AD of ammonia-containing substrates like animal manure and organic fractions of municipal solid waste are susceptible to ammonia inhibition. Besides ammonia and ammonium ions, inhibitory effects are also subjected to pH, temperature, reactor configuration, and type of inoculum [13]. Since free ammonia concentration is closely correlated with temperature, ammonia inhibition is more likely to occur at thermophilic temperatures than mesophilic ones. Methanogens are the most likely of the four anaerobic microbe types to grow intolerant of ammonia inhibition. Consequently, ammonia concentrations under 200 mg/L benefit anaerobic activities and do not hinder them [11].

1.4 Prospective Feedstocks for Biogas Production

All organic wastes originating from different sources in rural and urban areas can be used as a substrate for biogas production. These feedstocks differ significantly in composition, moisture content, homogeneity, and biodegradability. There is a wide range of organic feedstock such as easily degradable wastewater, complex high-solid waste, lignocellulosic residues, food waste, slaughterhouse waste, industrial waste, waste (seed cakes) generated from the production processes of ethanol and biodiesel, etc., as shown in Fig. 1.3.

Organic wastes, which contain carbohydrates, proteins, fats, oils, grease (called FOG), cellulose, and hemicelluloses as main components, are potential substrates for biomethane production. These groups of substrates theoretically lead to different

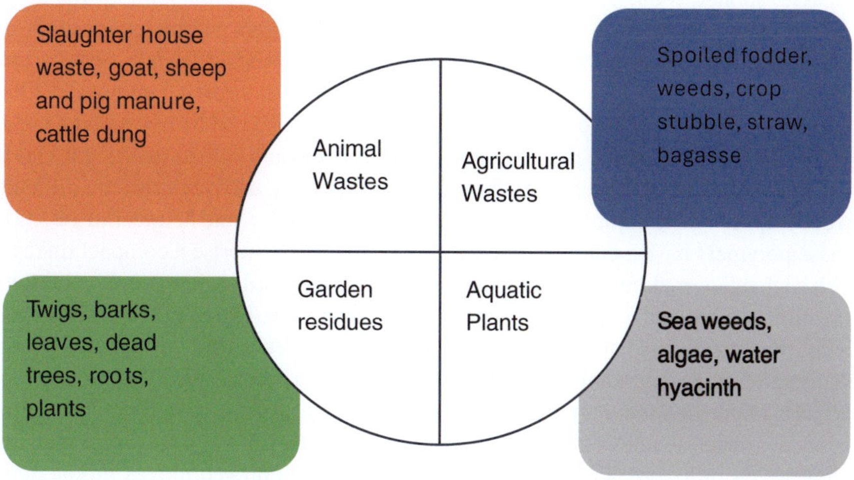

Fig. 1.3 Categories and types of feedstocks for anaerobic digestion

Table 1.3 Characteristics of some of the organic feedstocks [28]

Feedstock	Total solids (%)	Volatile solids (% of TS)	C:N ratio	Biogas yield (m^3 kg^{-1} VS)	Retention time (d)
Pig slurry	3–8	70–80	3–10	0.25–0.50	20–40
Cow slurry	5–12	75–85	6–20	0.20–0.30	20–30
Leaves	80	90	30–80	0.10–0.3	8–20
Wood shavings	80	95	511		–
Straw	70	70	90	0.15–0.35	10–50
Garden wastes	60–70	90	100–150	0.20–0.50	8–30
Fruit wastes	15–20	75	35	0.25–0.50	8–20
Food remains	10	80	2–18	0.50–0.60	10–20

methane and carbon dioxide compositions, as shown in the table. Fats produce the highest biogas, but they need a longer retention time. Protein and carbohydrates exhibit much faster conversion rates than fats [2]. The potential organic feedstock for biogas production can be classified into four broad categories: agricultural, animal, municipal, and industrial, with different characteristics as shown in Table 1.3.

1.4.1 Agricultural Wastes and Crop Residues

Worldwide, biomass is abundantly available in agricultural waste and crop residue. Crop processing yields residues, which are a potential feedstock for biogas production. Crop stubbles, straws and husks of various cereal crops, sugarcane trash, bagasse, spoiled fodder, weeds, coffee husks, oil cakes, etc., are some of the agricultural wastes used for biogas production. Table 1.3 shows the biogas generation potential from various agricultural wastes. The biogas production rate varies from one type of waste to the other. Wheat, rice, and bagasse straw contain over 20% lignin and have a C/N ratio of 120–150. These wastes need additives such as night soil with a low C/N ratio to maintain optimum C/N ratio and to increase biogas yield.

Lignocellulosic organic materials such as wood, wheat, and rice straw are difficult to decompose due to complex polymers like lignin, cellulose, and hemicellulose [32]. These polymers have a complex chemical structure and comprise a compact linkage with each other, making these lignocellulosic materials hard to degrade. In addition, low moisture content, the proportion of structural and non-structural carbohydrates, cellulose crystallinity, and the degree of bonding between carbohydrates and lignin also result in slow degradation [15]. Consequently, extra processing of these wastes in the form of pre-treatment is needed to make these materials easily degradable. Lignocellulose wastes contain a high C/N ratio, so to adjust the C/N ratio co-digestion with a lower C/N ratio, organic materials are practiced for improved biogas production. Other than crop residues, there are energy crops such as Napier grass and sorghum and woody crops (tree crops), which are appropriate for biogas production. These energy crops can be grown on low fertile land, requiring very little energy and costs for planting and harvesting. This property makes them an economical, feasible feedstock option for biogas production. Large volumes of stillage wastewater are produced from ethanol production from energy crops, which is also a potential feedstock for biogas production. Most agricultural wastes are co-digested with substrates like food waste, municipal biowaste collected from households, and energy crops to increase biogas yields.

Aquatic plants are also potential feedstocks for biogas production. These include micro- and macroalgae, water hyacinth, and seaweed. These feedstocks are known to have easily hydrolysable sugars and lesser lignin content [20]. Both microalgae and water hyacinth are used as feedstocks because of their higher gas yields. The relatively high and easily hydrolysable protein, lipid, and starch contents and the absence of lignin make microalgae an effective feedstock for biogas production.

1.4.2 Animal Manure

Animal manure is the largest organic waste material available almost everywhere. Generally, animal waste includes cattle manure, poultry manure, fishery waste, pig waste, horse dung, camel dung, and slaughterhouse waste. Cattle manure has been one of the oldest and most popular for anaerobic digestion. C/N around 30 makes it suitable for anaerobic digestion [23]. Its pH is generally neutral and has a high buffering capacity (ability to manage pH neutral). It contains a mix of naturally occurring bacteria essential for anaerobic degradation. It also delivers a mix of micronutrients, nutrients, and trace metals.

In rural India, cattle dung is a popular feedstock for anaerobic digestion. China, Korea, Taiwan, and the Philippines mostly use pig manure as feedstock, while India uses cattle dung for biogas production. Pig manure and poultry waste have a C/N ratio in the range of 12 to 15:1. So, to make it suitable for anaerobic digestion, co-digestion with carbon-rich feedstock is practiced. Co-digestion of two or more substrates provides a balance of macronutrients and micronutrients, optimum dilution, moisture balance, improved buffering capacity to the mixture, enhanced biodegradability, and expansion of the microbial community for an efficient digestion process [16]. Co-digestion leads to improved process stability and higher biogas yields. Blending nutrient-rich feedstocks such as food-processing waste and ethanol stillage with animal manure is a common practice for augmenting nutrient levels and providing the buffering capacity to maximize biogas production.

1.4.3 Municipal Waste

Rapid human expansion and increased urbanization and industrialization have led to substantial municipal solid waste (MSW). It is a heterogeneous waste, and its composition depends upon numerous factors such as regional differences, cultural practices, the extent of recycling, collection frequency, climate, etc. The management of MSW is a primary environmental concern across the world today. The range of prospective feedstocks under this category includes municipal wastewater, organic fractions of municipal solid waste, residual sludge, food remains, garden waste, and grass clippings [30].

Source-segregated organic fractions of municipal waste become a potential substrate for biogas generation. Sewage sludge from biological treatment plants and municipalities is generally found in the liquid state, and it is a dynamic material for biogas production due to its quick biodegradability. Moreover, the co-digestion of organic fractions of municipal solid waste with sewage sludge increases the nutrient level of feedstock and reduces the need for water.

Green waste is the waste from parks and gardens. This waste needs pulverization to provide homogeneous feedstock for biogas production. It may also contain woody lignocellulosic material, which requires physical/chemical pre-treatment to increase its biodegradability. Some of the municipal wastes are shown in the figure below.

a. Sewage
b. Green waste
c. Fruit and vegetable waste, food waste, kitchen waste, vegetable market waste
d. Municipal solid waste (OFMSW).

Food waste contributes a considerable fraction of MSW. With population and economic growth, the quantity of food waste generated has promptly increased and varies according to different eating habits. This class of waste includes kitchen waste, food remains from hotels and canteens, and waste from fruit and vegetable markets. The composition of food waste is different in proportion to vegetables, rice, eggs, meat, and other main components. These wastes are characterized by high moisture content (> 80%), VS content (> 95%), and high biodegradability [15].

1.4.4 Industrial Sources

Various industries and commercial entities exist that generate large volumes of organic wastewater that is friendly for anaerobic digestion. Some of the industrial sources are food/beverage processing, dairies, starch processing, breweries, sugarcane industry, slaughterhouses, fruit processing, pulp and paper, and textile industry. Food processing industries such as dairy processing, citrus processing, potato processing, vegetable canning, breweries, and sugar production industries process a large amount of raw agricultural material. These are perspective feedstocks for anaerobic digestion as well as co-substrates.

1.5 Biogas-Digested Slurry

During anaerobic digestion 25–30% of the total dry matter of organic waste gets converted into biogas, and 70–75% of the total solids content comes out as sludge, which is known as a digested slurry or biogas slurry [33]. Biogas-digested slurry is a nutrient-rich material that can be used as an organic fertilizer. It is good quality organic manure free from pathogens, weeds, and contaminants like heavy metals or infectious microorganisms. The high-quality digested slurry can be obtained by feeding biogas plants with high-quality substrates, such as source-separated organic waste. Any digested slurry's nutrient content reflects the feedstock's nature, which has undergone digestion.

Animals do not utilize all the nutrients in their feed during digestion in their bodies. This results in the excretion of high quantities of nitrogen, potassium, and phosphorus as excreta. These macro- and micronutrients are preserved during anaerobic digestion of animal dung. The subsequently digested manure from the digester thus constitutes a highly valuable product. The nitrogen in the cattle dung remains conserved, and about 20% of the total nitrogen is readily available for crops in the ammonical form when the digested slurry is utilized as fertilizer [31]. This helps to recycle nutrients through the use of biogas-digested slurry as a biofertilizer.

1.5.1 Benefits of Biogas-Digested Slurry

Biogas-digested slurry, a nutrient-rich material, can be directly used as a fertilizer on agricultural land. It has the potential to replace inorganic fertilizers and/or soil conditioners. It is rich in humus, and it improves the cation exchange capacity (CEC), aggregation, and water-holding capacity of the soils, stabilizes its humid content, and prevents the leaching of nutrients. It enhances agricultural productivity by providing a high content of soil nutrients and enzymes. Bioslurry is rich in the major plant nutrients (NPK) and micronutrients, which are essential for plant growth. It neither smells bad nor contains harmful parasites. The benefits of using digestate on agricultural land as an organic fertilizer in place of synthetic fertilizers include:

(1) **Enhanced soil quality**: The continuous use of chemical fertilizers in farms to increase crop yields results in deterioration in soil quality. Digested slurry enhances soil quality more effectively than artificial (synthetic) fertilizers [8]. Biogas-digested slurry has a high biological component that contributes to soil health and the formation of microscopic biofilms within soils. These components thus improve the capacity of the soil to retain more moisture and reduce the risk of wearing erosion and compaction.
(2) **Increased crop yields**: Even though the use of artificial, man-made fertilizers is largely intended to maximize crop yields. These artificial fertilizers can provide only particular nutrients for crops. The biogas-digested slurries provide the

same level of nutrition with the additional benefit of simultaneously providing nutrients to the soil.

(3) **Increased income**: Biogas digestate slurry as a fertilizer makes a strong economic sense in terms of effective nutrient value and economic returns as a fertilizer. Biogas slurry is a marketable product. The viability of biogas plants without slurry utilization is meager. Hence, in addition to the value of gas, the value of slurry as fertilizer has to be assessed in calculating the economics of the plant. Biogas slurry has more nutrients than Farmyard Manure (FYM) because the nutrients are not lost by evaporation due to exposure to the sun or leaching. However, there is an extra cost of applying digested slurry compared to artificial fertilizer (readily available in the market and cheap) due to transportation costs, storage, treatment and separation costs, etc. However, the nutrient benefits of the biogas digestate slurry and the sales returns overshadow the cost of processing the slurry. For example, a 200 m^3/day biogas plant requires 5 tons of cow dung as substrate; this produces approximately 1.25 tons of air-dried biogas digestate (manure). The average selling price of digestate is 4–10 cents per kg. This adds to the cost economics of the biogas plant.

1.5.2 Nutrient Value of Biogas-Digested Slurry

The state of the cattle dung before it is mixed with water, the breed and age of the cattle (buffalo, horse, elephant, etc.), the kind of feed given to the animals, their exposure to the sun, and the anaerobic digestion process caused by the activity of microorganisms, etc. Fresh dung typically has a dry matter content of 20%. If water is mixed in equal parts before feeding, the percentage will drop to 10%. In anaerobic digestion, roughly 30% of the organic content breaks down, decreasing the dry matter in the slurry to 7%. The fresh dung has roughly 1% nitrogen in it. Even though no new nitrogen is created during anaerobic digestion, a 30% loss of organic matter causes nitrogen concentration to increase to roughly 1.40%. Most nitrogen is transformed into ammonium under anaerobic circumstances, which makes it easily accessible for plant growth.

1.5.3 Techniques for Processing Biogas-Digested Slurry

The digested slurry can either be applied to the land as obtained from the biogas digester (called “whole”) or separated into solids and liquids, commonly denoted as “dried” (fiber) and “liquid” (liquor) form [12]. The whole digestate needs to be transported to the agricultural area. The transportation of the digested slurry to distant regions causes logistic difficulties because the digestate consists of 95% water. Digest slurry processing can primarily be required for volume reduction or to enhance quality. Dewatering of the digested slurry requires high investment and maintenance costs but drastically reduces transportation costs. Separating the digested slurry into solid and liquid fractions using mechanical, biological, or thermal equipment has several benefits:

- Facilitates separate storage and reduced volume of flowable (liquid) and stackable materials (solids).
- Enhanced application of separated liquid.
- Separation and increase of nutrients in different fractions in liquid and solid forms.

Many techniques have been developed for processing the digested slurry. These techniques are currently available with various degrees of technical development, specific features, and efficiencies. The effectiveness of the separation technique depends upon the nature of the digested slurry and the characteristics of residual particles. Commonly used methods of separating biogas-digested slurry into solid and liquid fractions are described below.

(i) **Screw Press**: It is often used for digestates having high fiber content. A screw press squeezes the digestate into a gradually reducing screw channel between a shaft and a screen mantle, pressing the liquid out. Different mesh sizes of the screen and working pressures can be used to vary the effectiveness of the process. Solid fraction (containing fibers) exits at the end of the separator. Screw presses have a separation efficiency between 10 and 40% depending on the feed characteristics and can achieve 30–38% dry solids in the separated solid fraction [21].

(ii) **Belt Filter Press**: A belt filter press is like a screw press but with a higher separation efficiency. It is used to produce cake rather than fiber at the output. The press consists of a closed loop of a textile belt wrapped around a cylinder with a decreasing diameter to press the water out. Firstly, the digestate obtained from the biogas plant is squeezed between tensioned belts. This dewaters the digestate by gravitation. Then, the digestate is pressed in between two filter belts by changing mechanical forces. Finally, a mechanical device removes the dry cake from the filter belt.

(iii) **Centrifuge**: It is a high-speed mechanical device that uses centrifugal force to separate small, fine, and colloidal particles from the digested slurry. Centrifuge machines are mainly used to separate the bulk amount of phosphorus with the solid fraction from the digested slurry. These usually have a separation efficiency of more than 95% of solids from the digested slurry. It can produce cake at the outlet with dry solids between 18 and 35% and liquor having less than 0.3% dry solids.

1.5.4 Utilizable Forms of Biogas-Digested Slurry

The digested slurry should be handled and applied correctly to the soil to avoid loss of ammoniacal nitrogen via volatilization. There are different modes of application of digested slurry in the field. It can be applied directly in the field using a container or discharged through an irrigation canal. It can also be applied to land as it is obtained from the digester "whole" or applied as solid or liquid fractions after processing. The utilizable forms of digestate fractions are explained below.

(i) **Whole Digested**

The whole digestate is the untreated digested effluent coming out from the digester. Normally, it has dry solids between 2 and 6% in a wet digester and between 18 and 23% in a dry digester, with varying concentrations of N, P, and K.

(ii) **Liquid (Liquor) Form**

The liquid form of digestate is also known as liquor due to its high water and low dry solid content. It has a relatively high value of N, P, K, in the order of 35:15:35 (kg/t); approximately half of the N is organic, and the other half is mineral (i.e., ammonium, NH_4). In the liquid state, it can be applied directly to the crops as a fertilizer by foliar spraying, buckets, tankers, or irrigation canal systems. Suppose the liquid digestate is applied to upright crops. In that case, it must be diluted with water in a ratio of 1:1.5–2.0 to avoid toxic effects on the plant growth and burning of the lower leaves of plants due to the presence of concentrated ammonia and soluble phosphorus. In addition, foliar spraying of slurry has many beneficial effects on leaves, crops, vegetables, and fruits concerning growth, quality, and disease resistance. Furthermore, some production systems using liquid digestate require ammonia conversion and/or supplementation for nutrient availability since the liquid form is lower in ammonia and nitrogen. To avoid ammonia (NH^+) loss, liquid slurry should be immediately used after it is transported to the field. Therefore, the storage of liquid slurry is a critical issue.

The digested sewage sludge from wastewater treatment plants typically has a high-water content. Therefore, unwatering is a prerequisite before spreading it in the fields (if the sludge comes under biomanure). The digested sludge generally has a high phosphorous content, which is valuable for plant nutrition. However, digested sludge can contain a high content of heavy metals, limiting its fertilizer use in agriculture. It can be mixed with structural and building materials such as sand and sawdust and is generally used as a filler material in the construction of roads, etc.

(iii) **Dried Form**

The dried form of the digestate refers to the grainy material separated from the whole digestate. The dried form mainly contains long fiber material and is generally used as a soil conditioner. Many farmers prefer the dry form of digested slurry for spreading in the fields as it is easier to transport than the liquid form. However, the dry digestate loses part of its nitrogen (particularly ammonia) after drying; therefore, the digestate's nutrient value is decreased. The digestate needs to be applied as soon as possible to avoid loss in nutritional value. Due to this, the dried form is the least efficient method of digestate application. The properties of the solid fraction are comparable to compost but are comparatively richer in lignocellulosic materials as lignin is not degraded during anaerobic digestion in the biogas plant.

The dried form of slurry is preferred to avoid tedious transportation and storage of liquid slurry. However, drying of the slurry results in some loss of the nitrogen and nutritive value of the slurry. Hence, this is the least efficient method.

1.5.5 Digested Compost

The above drawbacks can be overcome by utilizing slurry as compost. A high-quality product can be obtained by processing the digestate into compost. To compost the biogas-digested slurry, it is mixed with fresh organic waste material because the digested slurry itself contains less degradable material to sustain the composting process [4]. Composted form is the best way to overcome the transportation issue related to liquid digestate and the nutrient loss of the dried form. The composted digested slurry ensures a complete breakdown of the organic components and fixes mineral nitrogen on the humus, thereby reducing nitrogen loss. The composted digested slurry provides a good source of nutrients like nitrogen, phosphorus, and micronutrients like magnesium (Mg) and iron (Fe) and also acts as an additive to the composting process. The dry materials of the compost will absorb water in the digestate. As such, the manure will become moist and easily transported. Additionally, dry materials like litter and kitchen waste can be easily used. To minimize the loss of nutrients in the composted digestate, it should be taken to the crops only when required and mixed with the soil quickly. The composted form can also be used for algae production. Besides using slurry straight as fertilizer, it can also be used for algae production, mixed with animal feed for supplementary nutrient addition, and used in mushroom and fish production.

References

1. Abbasi T, Tauseef SM, Abbasi SA (2012) A brief history of anaerobic digestion and "biogas". In: Biogas energy. SpringerBriefs in Environmental Science, vol 2. Springer, New York. https://doi.org/10.1007/978-1-4614-1040-9_2
2. Achinas S, Achinas V, Euverink GJW (2017) A technological overview of biogas production from biowaste. Engineering 3(3):299–307. https://doi.org/10.1016/J.ENG.2017.03.002
3. Adekunle KF, Okolie JA (2015) A review of biochemical process of anaerobic digestion. Adv Biosci Biotechnol 06(03):205–212. https://doi.org/10.4236/abb.2015.63020
4. Al Seadi T, Drosg B, Fuchs W, Rutz D, Janssen R (2013) Biogas digestate quality and utilization. In: The Biogas handbook: science, production and applications, pp 267–301. https://doi.org/10.1533/9780857097415.2.267
5. Ali Shah F, Mahmood Q, Maroof Shah M, Pervez A, Ahmad Asad S (2014) Microbial ecology of anaerobic digesters: the key players of anaerobiosis. Sci World J 2014. https://doi.org/10.1155/2014/183752
6. Bharathiraja B, Sudharsana T, Jayamuthunagai J, Praveenkumar R, Chozhavendhan S, Iyyappan J (2018) Biogas production—a review on composition, fuel properties, feed stock and principles of anaerobic digestion. In: Renewable and sustainable energy reviews, vol 90. Elsevier, Amsterdam, pp 570–582. https://doi.org/10.1016/j.rser.2018.03.093

7. Bond T, Templeton MR (2011) History and future of domestic biogas plants in the developing world. Energy Sustain Dev 15(4):347–354. https://doi.org/10.1016/J.ESD.2011.09.003
8. Bonten LTC, Zwart KB, Rietra RPJJ, Postma R, De Haas MJG, Wageningen A, Wageningen UR (2014) Bio-slurry as fertilizer Is bio-slurry from household digesters a better fertilizer than manure? A literature review. www.wageningenUR.nl/en/alterra
9. Chandra R, Takeuchi H, Hasegawa T (2012) Methane production from lignocellulosic agricultural crop wastes: a review in context to second generation of biofuel production. Renew Sustain Energy Rev 16(3):1462–1476. https://doi.org/10.1016/J.RSER.2011.11.035
10. Chasnyk O, Sołowski G, Shkarupa O (2015) Historical, technical and economic aspects of biogas development: case of Poland and Ukraine. Renew Sustain Energy Rev 52:227–239. https://doi.org/10.1016/J.RSER.2015.07.122
11. Chen Y, Cheng JJ, Creamer KS (2008) Inhibition of anaerobic digestion process: a review. Biores Technol 99(10):4044–4064. https://doi.org/10.1016/J.BIORTECH.2007.01.057
12. Drosg B, Fuchs W, Al T, Madsen SM, Linke B (2015) Nutrient recovery by biogas digestate processing
13. Garcia ML, Angenent LT (2009) Interaction between temperature and ammonia in mesophilic digesters for animal waste treatment. Water Res 43(9):2373–2382. https://doi.org/10.1016/J.WATRES.2009.02.036
14. Gerardi MH (2003) The microbiology of anaerobic digestors. In: The microbiology of anaerobic digesters. Wiley. https://doi.org/10.1002/0471468967.fmatter
15. Gunaseelan VN (1997) Anaerobic digestion of biomass for methane production: a review. Biomass Bioenerg 13(1–2):83–114. https://doi.org/10.1016/S0961-9534(97)00020-2
16. Hagos K, Zong J, Li D, Liu C, Lu X (2017) Anaerobic co-digestion process for biogas production: progress, challenges and perspectives. Renew Sustain Energy Rev 76:1485–1496. https://doi.org/10.1016/J.RSER.2016.11.184
17. Jain S, Jain S, Wolf IT, Lee J, Tong YW (2015) A comprehensive review on operating parameters and different pretreatment methodologies for anaerobic digestion of municipal solid waste. Renew Sustain Energy Rev 52:142–154. https://doi.org/10.1016/J.RSER.2015.07.091
18. Kougias PG, Angelidaki I (2018) Biogas and its opportunities—a review. Front Environ Sci Eng 12(3). https://doi.org/10.1007/s11783-018-1037-8
19. Kumar KV, Sridevi V, Rani K, Sakunthala M, Kumar CS (2013) A review on production of biogas, fundamentals, applications & its recent enhancing techniques
20. Kwaku Armah E, Kweinor Tetteh E, Boafo Boamah B (2017) Overview of biogas production from different feedstocks. Int J Sci Res Publ 7(12). www.ijsrp.org
21. Lukehurst CT, Frost P, Al Seadi T (2010) Utilisation of digestate from biogas plants as biofertiliser
22. Mao C, Feng Y, Wang X, Ren G (2015) Review on research achievements of biogas from anaerobic digestion. Renew Sustain Energy Rev 45:540–555. https://doi.org/10.1016/J.RSER.2015.02.032
23. Monnet F (2003) An introduction to anaerobic digestion of organic wastes. A report by Remade, Scotland. http://www.biogasmax.co.uk/media/introanaerobicdigestion__073323000_1011_24042007.pdf
24. Rajagopal R, Massé DI, Singh G (2013) A critical review on inhibition of anaerobic digestion process by excess ammonia. Biores Technol 143:632–641. https://doi.org/10.1016/J.BIORTECH.2013.06.030
25. Schon M (2009) Numerical modelling of anaerobic digestion processes in agricultural biogas plants
26. Sher F, Smječanin N, Hrnjić H, Karadža A, Omanović R, Šehović E, Sulejmanović J (2024) Emerging technologies for biogas production: a critical review on recent progress, challenges and future perspectives. Process Saf Environ Prot 188:834–859. https://doi.org/10.1016/J.PSEP.2024.05.138
27. Siddique MNI, Wahid ZA (2018) Achievements and perspectives of anaerobic co-digestion: a review. J Clean Prod 194:359–371. https://doi.org/10.1016/J.JCLEPRO.2018.05.155
28. Steffen R, Szolar O, Braun R (1998) Feedstocks for anaerobic digestion

29. Stürmer B (2017) Biogas—part of Austria's future energy supply or political experiment? Renew Sustain Energy Rev 79:525–532. https://doi.org/10.1016/J.RSER.2017.05.106
30. Tyagi VK, Fdez-Güelfo LA, Zhou Y, Álvarez-Gallego CJ, Garcia LIR, Ng WJ (2018) Anaerobic co-digestion of organic fraction of municipal solid waste (OFMSW): progress and challenges. Renew Sustain Energy Rev 93:380–399. https://doi.org/10.1016/J.RSER.2018.05.051
31. Wagaw K (2016) Characterization and Utilization of bioslury from anaerobic digester for fertilizer in crop production. J Fertilizers Pesticides 7(12). https://doi.org/10.4172/2471-2728.1000169
32. Ward AJ, Hobbs PJ, Holliman PJ, Jones DL (2008) Optimisation of the anaerobic digestion of agricultural resources. Bioresour Technol 99(17):7928–7940. https://doi.org/10.1016/j.biortech.2008.02.044
33. Warnars L, Oppenoorth H (2014) Bioslurry: a supreme fertiliser : a study on bioslurry results and uses. Hivos
34. Wellinger A (1999) Process design of agricultural digesters
35. Westerholm M, Schnürer A (2019) Microbial responses to different operating practices for biogas production systems. www.intechopen.com
36. Wu D, Li L, Zhao X, Peng Y, Yang P, Peng X (2019) Anaerobic digestion: a review on process monitoring. Renew Sustain Energy Rev 103:1–12. https://doi.org/10.1016/J.RSER.2018.12.039
37. Yadav M, Joshi C, Paritosh K, Thakur J, Pareek N, Masakapalli SK, Vivekanand V (2022) Organic waste conversion through anaerobic digestion: a critical insight into the metabolic pathways and microbial interactions. Metab Eng 69:323–337. https://doi.org/10.1016/J.YMBEN.2021.11.014
38. Yadav M, Joshi C, Paritosh K, Thakur J, Pareek N, Masakapalli SK, Vivekanand V (2022a) Organic waste conversion through anaerobic digestion: A critical insight into the metabolic pathways and microbial interactions. In: Metabolic engineering, vol 69. Academic, pp 323–337. https://doi.org/10.1016/j.ymben.2021.11.014
39. Yenigün O, Demirel B (2013) Ammonia inhibition in anaerobic digestion: a review. Process Biochem 48(5–6):901–911. https://doi.org/10.1016/J.PROCBIO.2013.04.012
40. Zhang L, Loh KC, Zhang J (2019) Enhanced biogas production from anaerobic digestion of solid organic wastes: current status and prospects. Bioresour Technol Rep 5:280–296. https://doi.org/10.1016/J.BITEB.2018.07.005
41. Zhang C, Su H, Baeyens J, Tan T (2014) Reviewing the anaerobic digestion of food waste for biogas production. Renew Sustain Energy Rev 38:383–392. https://doi.org/10.1016/J.RSER.2014.05.038

Chapter 2
Biogas: Quality, Standards, and Regulations for Utilization

2.1 Basics of Biogas

Biogas produced from the anaerobic decomposition of the organic fraction of all wastes is a renewable, clean, and environmentally friendly fuel. It is acknowledged worldwide as one of the foremost technologies for producing green gas and a valuable fertilizer from digestible biomass. Methane (CH_4) and carbon dioxide (CO_2) make up the majority of the gas extracted from biogas, with trace amounts of nitrogen (N_2), hydrogen (H_2), oxygen (O_2), water vapor (H_2O), hydrogen sulfide (H_2S), siloxanes, and saturated hydrocarbons (ethane, propane) based on the type of feedstock used.

The typical constituents of biogas are shown in Table 2.1, and the exact proportions depend upon the type and chemical composition of the substrate used for biogas production, the production conditions, and processing practices.

The density of biogas is almost 20% less than the air. Its ignition temperature is 650–750 °C, and flammability limits on the air of 8–18%v/v. The calorific value of raw biogas varies from 20 to 25 MJ/Nm^3 depending upon the composition. At 50% CH_4 composition, the heating value is 21 MJ/Nm^3 while its density is 1.22 kg/Nm^3 (comparable to air, 1.29 kg/Nm^3). Typically, biogas has a specific gravity of 0.86 and a Wobbe number ($\approx$ 27 MJ/ Nm^3). With a 60% composition of CH_4, 1 m^3 of raw biogas produces 21.5 MJ (5.97 kWh) heat on burning, whereas 1 m^3 of pure CH_4 produces 35.8 MJ (9.94 kWh) heat at the standard temperature and pressure conditions [4]. As the methane content of raw biogas is the deciding factor of its calorific value, it should be considerably high to provide high energy content and to make the biogas suitable for basic energy applications like cooking and electricity production. Table 2.2 provides the calorific values of biogas with various fuels used in power applications.

V. K. Vijay et al., *Compressed Biogas (CBG)*, Green Energy and Technology,
https://doi.org/10.1007/978-981-96-5697-4_2

Table 2.1 Composition of biogas produced from different feedstocks [3, 18, 21]

Compound	Combustible/ non-combustible	Chemical formula	Agriculture/ animal waste	Municipal waste	Wastewater	Landfill
Methane	Combustible	CH_4	50–75	50–60	60–75	40–70
Carbon dioxide	Non-combustible	CO_2	30–50	35–38	20–45	15–40
Nitrogen	Non-combustible	N_2	1–2	0–5	0–2	1–20
Hydrogen	Combustible	H_2	–	–	–	0–5
Oxygen	Non-combustible	O_2	0.5–1	0–1	0.5–1	0–5
Water vapor	Non-combustible	H_2O	100% saturated at biogas digester outlet	100% saturated at biogas digester outlet	100% saturated at biogas digester outlet	100% saturated at biogas digester outlet
Hydrogen culfide	Combustible	H_2S	3–7000	70–650	50–3000	0–20,000
Siloxanes	Non-combustible		< 0.4		1.5–15	0.1–4
Halogenated hydrocarbons	Non-combustible			100–800		1–2900

Table 2.2 Calorific and energy values of various fuels [11, 12, 17]

Fuel	Calorific value [Kcal]	Energy [kWh]
1 Nm^3 raw biogas	5200	5.97
1 Nm^3 upgraded biogas (97% methane)	11,942	9.67
1 Nm^3 natural gas	8600	11.0
1 L petrol	11,200	9.06
1 L diesel	10,700	9.8
1 L LPG	10,500	12.2

2.2 Major Constituents of Biogas

Methane makes up 55–70% of biogas, whereas carbon dioxide makes up 30–45%. Additional gases (contaminants) include ammonia (0–100 ppmv), oxygen (0–3%), water (1–3%), hydrocarbons (HC) (0–200 mg m^3), nitrogen (0–15%), and siloxanes (0–41 mg Si m^3). Biogas primarily consists of methane. It is an odorless, non-toxic, and extremely flammable part of biogas. Methane does not produce any harmful gases during combustion, only carbon dioxide and water, despite being a greenhouse gas that is 25 times more harmful than carbon dioxide when used as a fuel. The quantity of methane in biogas is a function of the digestion temperature, as at low digestion temperatures, methane content is high, but the yield is less. The second significant portion of biogas is carbon dioxide. It is the non-combustible part of biogas and does not add to the calorific value of biogas [2]. Its inclusion raises the biogas's methane number and anti-knock qualities while lowering the total heating value. Carbon dioxide and methane have no color or smell.

Anaerobic decomposition produces hydrogen sulfide, which gives raw biogas its characteristic odor. The amount of sulfur in the feedstock and the digester's pH both affect how much H_2S is produced. It is formed due to the conversion of elemental sulfur of feedstock during the acidogenesis phase in anaerobic digestion. Due to its corroding property, H_2S is toxic and poses a significant risk to human health and the piping system. Biogas production in anaerobic digestion is a moist or water-based process, and the presence of water vapor as a constituent can cause corrosion separately and with other constituents, like H_2S, in pipelines on condensation. The presence of nitrogen and oxygen in biogas shows the improper sealing or trapping of environmental air in the digester at the time of its closing.

2.3 Cleaning and Upgradation of Biogas

Raw biogas from the digester must be cleaned from components like water vapor, hydrogen sulfide, siloxanes, dust, etc., to safeguard the mechanical components from corrosion and reach specific quality standards for efficient utilization. Biogas requires treatment in terms of cleaning and upgrading for two main reasons.

(a) To reduce the chance of damaging and corroding the process equipment.
(b) To increase the product gas's energy density (calorific value).

Biogas treatment is done in two steps: cleaning and upgradation. In the cleaning step, harsh and corrosive components, mainly hydrogen sulfide and moisture, are removed, while in the upgradation step, primarily carbon dioxide is removed from biogas to increase its energy density. These treatment steps increase the heating/calorific value and reduce the adequate storage volume of the cylinders. Biogas cleaning can be carried out as an individual unit operation (dedicated only to biogas cleaning) depending upon the end use of raw biogas; for some applications that require stringent gas quality, both cleaning and upgrading treatment steps are applied. Usually, biogas is cleaned before the upgrading process, but in some processes, such as the water scrubbing process, both cleaning and upgrading processes can be carried out simultaneously in a single operation. The effect of various constituents present in raw biogas according to their applications is described in Table 2.3.

2.3.1 Technologies for Biogas Cleaning and Upgrading

Biogas cleaning and upgrading is a gas separation problem. The separation processes are based on the differences in thermodynamic, transport, or molecular properties of the gaseous components in biogas. Thermodynamic and transport properties exploit

Table 2.3 Constituents of biogas, sources, and impacts [7, 9, 15, 16, 20]

Component	Source	Impact
Carbon dioxide	• Carbon from organic waste	• Reduces the biogas's heating value • Raises the engine's anti-knock characteristics and methane number • Produces carbonic acid when there is moisture present, which leads to corrosion • Increases energy density of biogas and reduces the cylinders' adequate storage volume
Hydrogen sulfide	• Proteins, sulfur-containing compounds present in organic waste	• Corrode equipment and piping system by forming acids with condensed water vapor
Water vapor	• Moisture present in organic waste	• Equipment and pipe systems corroded as a result of the CO_2 and H_2S reactions, which formed acids
Siloxane	• Cosmetics, anti-foaming and washing agents	• Causes abrasion and damage to engine parts
Nitrogen	• Air	• Brings down the heating value • Enhances the anti-knock property of the engine
Dust		• Blocks nozzles and engine parts

solubility, adsorption capacity, vapor pressure, boiling point, and diffusivity of the gaseous components. The discrepancies between the molecules' quadrupole, dipole, kinetic diameter, and polarizability are the focus of molecular properties. Based on these properties of the gaseous components, the processes that separate the gases are primarily categorized into sorption and separation [1, 6]. The terms "sorption" and "adsorption" refer to the processes of atoms, molecules, or even ions adhering to materials' surfaces and forming an adsorbate layer on the surface of the adsorbent, and of absorption (atoms, molecules, or even ions absorbed either physically or chemically into a solvent). Membrane technology and cryogenic techniques that use membranes based on vapor pressure and boiling temperatures, respectively, to separate the contents between two fractions are included in the term "separation." Commercially accessible methods for cleaning and upgrading biogas include membrane separation, chemical scrubbing, pressure swing adsorption, water scrubbing, and cryogenic separation. These technologies are based on the previously mentioned approach.

2.4 Technical Demands for the Utilization of Biogas

Upgraded biogas is lagging in adoption as a fuel due to the unavailability of a technical standard on an international level. Very few countries have national standards for upgrading biogas as a substitute for natural gas. Gas quality standards are defined by the basic requirements for gas in terms of composition and combustion properties,

Table 2.4 Natural gas/biomethane standards of various countries [2, 5, 7, 8, 9, 13, 14, 19]

Country	Austria	France	Germany	Netherlands	Sweden	Switzerland	British Columbia	California	Michigan	India
Calorific upper value (MJ/Nm^3)	38.5–46.1	38.5–46.1 (H gas) 34.2–37.8 (L gas)	30.2–47.2	31.6–38.7		38.5–47.2		36.9–42.8		
Wobbe index (MJ/Nm^3)	47.9–56.5 (upper)	48.2–56.5 (H gas) 42.5–46.8 (L gas upper)	46.1–56.5 (H gas) 42.5–46.8 (L gas upper)	43.6–44.41 (upper)	44.7–46.4 (Type A)[a] 43.9–47.3 (Type B)[a] (lower)	47.9–46.1 (unlimited injection)	48.2–56.5 (H gas) L: 42.5–46.8 (L gas upper)	47.6–51.6 (upper)		
Qualities										
CH_4 (%)[b]	> 96			> 80	97 ± 1 (Type A) 97 ± 2 (Type B)	> 96 (unlimited injection) > 50 (limited injection)	> 95.5		> 93.5	> 90
CO_2 (%)[b]	< 2	< 2.5[c]	< 6	< 6 (< 10–10.3 for regional grid)	< 3	< 6 (unlimited injection) < 4 (limited injection)	< 2	3	< 2	< 4
$CO_2 + O_2 + N_2$ (vol. %)					< 4 (type A)[a] < 5 (type B)[a]					< 10
H_2 (%)[b]	< 4	< 6	< 5	< 12	< 0.5	< 4		0.1		

(continued)

Table 2.4 (continued)

Country	Austria	France	Germany	Netherlands	Sweden	Switzerland	British Columbia	California	Michigan	India
O_2 (%)[b]	< 0.5	< 0.01[c]	< 3	< 0.5	< 1	< 0.5	< 0.2	< 0.2	< 3	< 0.5
CO(%)[b]		< 2		< 1		< 0.5				
Total inert (vol %) (CO_2, N_2, O_2, CO, H_2)								4		
Water content Max. (mg/Nm^3)				< 32	< 32		< 65		No Condensation	16 mg/m3
Total sulfur (mg/m^3)	< 10	< 30	< 30[d]	< 45	< 23	< 30		265		
H_2S	< 5 mg/m^3	< 5 mg/m^3 (H_2S + COS)	< 5 mg/m^3	< 5 mg/m^3	< 10 ppm = < 15.2 mg/m^3	< 5 mg/m^3	4.3 ppm	88 mg/m^3	4.1 ppm	< 30.3 mg/m^3

[a] First, there are two types of biogas fuels: type A is for vehicles with engines that do not have lambda control, and type B is for vehicles that do
[b] The symbol @ denotes volume percentage or moles percentage
[c] France permits some degree of parameter flexibility. Under some circumstances, the concentration of CO_2 and oxygen can be raised to 11.3% and 3.3%, respectively
[d] Exclusive odorization

transport, and environmental emission norms. The technical demands applicable to upgraded biogas (biomethane) can be stated as gas quality and interchangeability. The standards for upgraded biogas have been formulated to ensure its compatibility with natural gas so that it can easily be injected into the natural gas grid and utilized in vehicles. The standards facilitate not only the uniform tuning of upgraded equipment (upgrading units, compressors, storage vessels, etc.) to avoid corrosion and damage to the parts but also the adoption of upgraded biogas in natural gas-operated vehicles. Some countries have formulated standards or technical specifications for utilizing the upgraded biogas as a substitute for natural gas. Standards for upgraded gas quality and interchangeability are formulated to ensure the same quality of gas around the world. An established and stable quality leads to favorable situations, i.e., equivalent settings for biomethane plants, analysis equipment, and eventually, a decrease in investments and operating costs. Permission procedures for injecting upgraded biomethane into the already established natural gas grid will be significantly rationalized after the fulfillment of gas quality, and it motivates biogas producers and local stakeholders to establish such projects.

(i) ***Gas Quality***

The gas quality in the technical standard defines the compositional and combustion properties of the gas. Raw biogas contains various types of impurities that can result in serious problems for the pieces of equipment as well as for the environment if not separated from biogas. For example, impurities such as moisture, hydrogen sulfide, and carbon dioxide, if present even in trivial quantities can have a corrosive effect on the equipment. These impurities can also degrade the piping and end-use applications and equipment. The gas compositional characteristics, as defined in the standards, will outline appropriate limits on the potentially damaging constituents present in biogas.

(ii) ***Interchangeability***

Interchangeability is known as the capability to replace one gaseous fuel with others without pertinently altering safety, effectiveness, performance, and environmental emission norms. Interchangeability can also be described with quantifiable procedures broad and widespread application of the gas. Calorific value can be treated as a criterion, but it is not a decent meter of interchangeability. Upgraded biogas is a benign substitute for natural gas. Biomethane can be regarded as a cleaner (contains fewer contaminants) than conventional natural gas.

The interchangeability of fuel gases, such as natural gas, is also measured by the Wobbe Index (WI) or Wobbe number, which is frequently utilized in the requirements for gas supply and transport services. The ratio of the gas's square root relative density to its gross heating value is called the Wobbe number, which is a fundamental interchangeability index for gases.

$$\text{Wobbe Index, WI} = \frac{\text{HHV}}{\sqrt{\rho_{\text{gas}}/\rho_{\text{air}}}}$$

where

HHV higher heating value of the gas (KJ/kg)
ρ_{gas} density of gas
ρ_{air} density of air.

The quality of gas in the commercial sector is determined by its heating value. For natural gas, the Wobbe index is used as a specific measure of its heating value. Additionally, the methane number (MN) is an important parameter for assessing gas quality, particularly for natural gas-fueled vehicle internal combustion engines. Typically, MN 100 represents 100% purified methane, while MN 0 represents 100% purified hydrogen.

2.5 Biomethane Standards in Different Countries

The implementation of standards and regulations in a country enables the potential injection of biomethane as a substitute or supplementary gas in transport and distribution networks. Biomethane standards offer guidance on biomethane composition and its use across various applications. Europe has taken the lead in developing and promoting the biomethane market, utilizing raw biogas as a sustainable energy source for powering diverse systems. The majority of European nations have set requirements for improved biogas injection into the natural gas system. These nations include Sweden, Norway, Denmark, Austria, Germany, Switzerland, Netherlands, Luxembourg, Finland, and the UK. While Canada and the USA also utilize upgraded biogas with existing natural gas infrastructure, national standards are yet to be developed in these countries. France and the Netherlands have stringent regulations for biomethane, with some flexibility, whereas Germany, Sweden, and Switzerland have more lenient standards. Standards for the use of upgraded biogas in stationary and mobile engines have been developed by the Bureau of Indian Standards. These standards can be implemented after obtaining the necessary licenses and permits to use biomethane as a transportation fuel. These standards provide thorough methods and recommendations for the composition of biomethane and its loading into CNG cylinders. They also detail the requirements for the use of biomethane in stationary engines, automotive applications, heating applications, and supply through the piped network.

As of right now, there is no global standard for either injecting improved biogas into the natural gas system or using it as vehicle fuel. Different nations have implemented national rules, standards, and processes that vary globally in terms of parameters, values, and measuring units (vol. %, mol.%, ppm). There is agreement on the necessary impurities that must be taken into account, even though there is disagreement over permissible limits for trace amounts of biogas constituents. When evaluating the quality of gas, factors such as the quantity of methane, the Wobbe index, water, CO_2, O_2, H_2, sulfur compounds, and heating value are crucial and usually lie

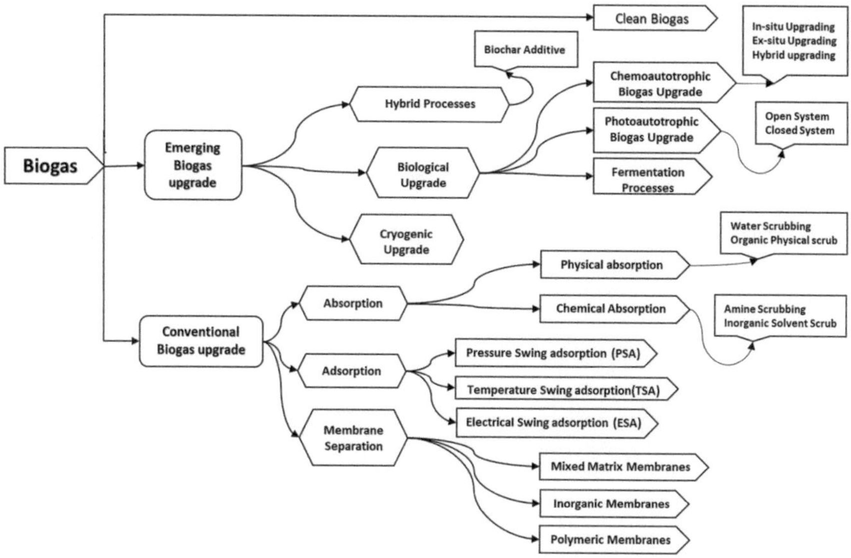

Fig. 2.1 Technologies for biogas cleaning and upgrading. Reproduced with Permission from Elsevier [10]

within the same range across national borders. Some countries monitor minor and uncertain elements of biogas based on the feedstocks used in anaerobic digestion. Sulfur is a highly debated parameter in the automotive industry, and it should be maintained at low levels to ensure system durability, with the suggested range for sulfur in normal grade biogas being 10–30 mg/m^3. Similarly, siloxanes can adhere to various engine parts, leading to abrasion and blockages, and although there is no strict limit for siloxanes in biogas, a range of 0.3–1 mg per cubic meter of biogas is considered acceptable.

In addition, if the gas is transported outside of a dry system or in close proximity to subterranean storage systems, strict limits on the amount of oxygen and sulfur must be set. To keep every component of the system safe under all pressure and temperature circumstances, moisture must be removed from the biogas. Three classes of water dew points − 10, − 20, and − 30 °C are considered at 200 bars of pressure to enable its adaptation to the environment on a national scale. The standards generally allow a maximum of 32 mg of moisture content in 1 m^3 of upgraded biogas for grid injection.

References

1. Adnan AI, Ong MY, Nomanbhay S, Chew KW, Show PL (2019) Technologies for biogas upgrading to biomethane: a review. Bioengineering 6(4). https://doi.org/10.3390/bioengineering6040092

2. Awe OW, Zhao Y, Nzihou A, Minh DP, Lyczko N (2017) A review of biogas utilisation, purification and upgrading technologies. Waste Biomass Valorization 8(2):267–283. https://doi.org/10.1007/s12649-016-9826-4
3. Baena-Moreno FM, Rodríguez-Galán M, Vega F, Vilches LF, Navarrete B (2019) Review: recent advances in biogas purifying technologies. Int J Green Energy 16(5):401–412. https://doi.org/10.1080/15435075.2019.1572610
4. Bharathiraja B, Sudharsana T, Jayamuthunagai J, Praveenkumar R, Chozhavendhan S, Iyyappan J (2018) Biogas production—a review on composition, fuel properties, feed stock and principles of anaerobic digestion. Renew Sustain Energy Rev 90:570–582. https://doi.org/10.1016/j.rser.2018.03.093
5. BIS, Biomethane Standards (2016) IS 16087. https://archive.org/details/gov.in.is.16087.2013. Accessed July 2015
6. Chaemchuen S, Zhou K, Verpoort F (2016) From biogas to biofuel: materials used for biogas cleaning to biomethane. ChemBioEng Rev 3:250–265. https://doi.org/10.1002/cben.201600016
7. Kadam R, Panwar NL (2017) Recent advancement in biogas enrichment and its applications. Renew Sustain Energy Rev 73:892–903. https://doi.org/10.1016/j.rser.2017.01.167
8. Kapoor R, Ghosh P, Kumar M, Vijay VK (2019) Evaluation of biogas upgrading technologies and future perspectives: a review. Environ Sci Pollut Res 26(12):11631–11661. https://doi.org/10.1007/s11356-019-04767-1
9. Khan IU, Othman MHD, Hashim H, Matsuura T, Ismail AF, Rezaei-Dasht Arzhandi M, Azlee IW (2017) Biogas as a renewable energy fuel—a review of biogas upgrading, utilisation and storage. Energy Convers Manage 150:277–294. https://doi.org/10.1016/j.enconman.2017.08.035
10. Khan MU, Lee JTE, Bashir MA, Dissanayake PD, Ok YS, Tong YW, Shariati MA, Wu S, Ahring BK (2021) Current status of biogas upgrading for direct biomethane use: a review. Renew Sustain Energy Rev 149. https://doi.org/10.1016/j.rser.2021.111343
11. Kumar A, Flynn P, Sokhansanj S (2008) Biopower generation from mountain pine infested wood in Canada: an economical opportunity for greenhouse gas mitigation. Renew Energy 33(6):1354–1363. https://doi.org/10.1016/J.RENENE.2007.07.008
12. Minde GP, Magdum SS, Kalyanraman V (2013) Biogas as a sustainable alternative for current energy need of India. J Sustain Energy Environ 4
13. Muñoz R, Meier L, Diaz I, Jeison D (2015) A review on the state-of-the-art of physical/chemical and biological technologies for biogas upgrading. Rev Environ Sci Biotechnol 14(4):727–759. Springer, Netherlands. https://doi.org/10.1007/s11157-015-9379-1
14. Petersson A, Wellinger A (2009) Biogas upgrading technologies-developments and innovations. www.iea-biogas.net
15. Ryckebosch E, Drouillon M, Vervaeren H (2011) Techniques for transformation of biogas to biomethane. Biomass Bioenergy 35(5):1633–1645. https://doi.org/10.1016/j.biombioe.2011.02.033
16. Singhal S, Agarwal S, Arora S, Sharma P, Singhal N (2017) Upgrading techniques for transformation of biogas to bio-CNG: a review. Int J Energy Res 41(12):1657–1669. https://doi.org/10.1002/er.3719
17. Stelte W, Holm JK, Sanadi AR, Barsberg S, Ahrenfeldt J, Henriksen UB (2011) Fuel pellets from biomass: the importance of the pelletizing pressure and its dependency on the processing conditions. Fuel 90(11):3285–3290. https://doi.org/10.1016/J.FUEL.2011.05.011
18. Sun Q, Li H, Yan J, Liu L, Yu Z, Yu X (2015) Selection of appropriate biogas upgrading technology—a review of biogas cleaning, upgrading and utilisation. Renew Sustain Energy Rev 51:521–532. https://doi.org/10.1016/J.RSER.2015.06.029
19. Svensson M (2014) European workshop biomethane 140311 Svensson-Biomethane Standards 1. www.socalgas.com/documents/business/Rule30_BiomethaneGuidance.pdf

20. Yang L, Ge X, Wan C, Yu F, Li Y (2014) Progress and perspectives in converting biogas to transportation fuels. Renew Sustain Energy Rev 40:1133–1152. https://doi.org/10.1016/j.rser.2014.08.008
21. Yentekakis IV, Goula G (2017) Biogas management: advanced utilization for production of renewable energy and added-value chemicals. Front Environ Sci 5(FEB). https://doi.org/10.3389/fenvs.2017.00007

Chapter 3
Technologies for Biogas Cleaning

3.1 Introduction

Raw biogas consists of several impurities, like CO_2, CO_2, siloxanes, NH_3, oxygen, moisture, dust, and several other components; the expulsion of these compounds is necessary to obtain biogas that meets the specifications. Various technologies are available through which specific quality biogas could be attained commercially. There are many ways to remove single or multiple compounds from biogas. These processes are similar to the natural gas treatment methods and the processes involved in controlling air pollution. However, research is still underway to further develop and optimize thesc techniques for their application in biogas upgrades. Every technology has unique advantages and disadvantages. However, no technology is a perfect fit for every circumstance. Gas with the desired valuable quality for a particular application depends on the proper cleaning technology.

3.2 Cleaning of Biogas

The biogas enrichment process starts with "biogas cleaning." This involves the rejection of toxic constituents (such as particulate matter, CO_2, H_2O, CO_2, NH_3, siloxanes, and other volatile organic compounds (VOCs)). To make biogas eligible for application in automobiles, CHP engines, and industrial boilers or introduce it in existing cities/town grid networks of natural gas, the removal of contaminants is essential. This is also necessary because harmful components can deteriorate the quality of downstream equipment due to corrosion; harmful gases may also be responsible for abrasion [23]. Table 3.1 shows the requirements for removing water vapor, CO_2, and CO_2 depending upon applications.

V. K. Vijay et al., *Compressed Biogas (CBG)*, Green Energy and Technology,
https://doi.org/10.1007/978-981-96-5697-4_3

Table 3.1 Requirements of removal of water vapor, CO_2, and CO_2 depending upon applications [7, 8]

Application	H_2O (water vapor)	CO_2	CO_2, siloxanes (ppm)
Boiler	Not required	Not required	Compulsory
Gas stoves	Not required	Not required	Compulsory
Combined heat and power engines	Required	Not required	Compulsory
Turbines	Not required	Not required	Compulsory
Micro-turbines	Not required	Not required	Compulsory
Fuel cells	Not required	Not required	Compulsory
Vehicular fuel	Compulsory	Compulsory	Compulsory
Natural gas grid network	Compulsory	Compulsory	Compulsory

The established gas purification processes for the separation of gases like CO_2, CO_2, and CH_4 are designed keeping in view the molecular properties like dissimilarities in kinetic diameters of the gases, polarizability, quadrupole, and dipole moments of the molecules of the gases). Alternatively, the thermodynamic properties (specific heat, heat capacity, entropy, and enthalpy) and transport properties (boiling points, solubility, adsorption capacity, and diffusivity) of the gaseous compounds in raw biogas [22]. The methods used for biogas treatment (cleaning and upgrading) are membrane (high pressure, low pressure), adsorption (pressure swing adsorption), cryogenic separation, and absorption (physical or chemical absorption, pressured water washing). Table 3.2 gives an overview of the consequences and the processing techniques for removing biogas components like water vapor, hydrogen sulfide, and carbon dioxide, which are treated as contaminants.

3.3 Water Vapor

Biogas that leaves the digester is saturated with water vapor in the 0–10% range. However, its content depends on the temperature; at 35 °C, biogas has a water vapor concentration of approximately 5%. This makes it necessary to dry biogas needs to be used in any application. Removing water vapor is essential as it reacts with other gaseous constituents, such as hydrogen sulfide or halogenated compounds, to produce corrosive acids that would corrode and damage equipment parts and pipes. Different standards have been formulated to achieve a correct gas composition for different applications, and water vapor removal is an essential and primary step to achieve them. The methods for removing water vapor are generally categorized into physical and chemical separation (Table 3.3).

Table 3.2 Consequences and the processing techniques for the removal of water vapor, CO_2, and CO_2 [13, 14, 23]

Biogas component	Consequence	Process	Technique
Water vapor	• Corrosion in equipment • The build-up of water droplets in pipes of the equipment • Condensation/freezing of piping system and nozzles	Cleaning	• Condensation • Adsorption • Refrigeration • Absorption
Hydrogen sulfide	• Corrosion in types of equipment and piping system	Cleaning	• Absorption with liquids • Biological doping with air/O2 • Biological infiltration • Membrane separation • Adsorption
Carbon dioxide	• Reduction in calorific value	Upgrading	• Adsorption (pressure swing adsorption) • Absorption (physical, chemical, or pressurized water scrubbing) • Membrane (high pressure, low pressure) • Cryogenic separation

3.3.1 Physical Separation Methods

Water condensation is the foundation for physical drying. Cooling at atmospheric pressure may eliminate water vapor from biogas most easily. Since this approach can only lower the dew point to 0.5 °C, it is less effective in removing water vapor and may cause operational issues like the surface of the heat exchanger freezing. Thus, it is usually used as a pre-treatment step before any other technique. To further decrease the dew point, biogas pressure has to be increased before cooling and then expanded to the preferred pressures. The condensed water drops are entrapped and disposed of as wastewater through the outlet pipe.

Physical separation techniques for the removal of water vapor from biogas include:

(i) Demisters—water vapor is removed from raw biogas using a wired mesh (micropores 0.5–2 nm). A dew point of 2–20 °C (atmospheric pressure) can be obtained using demisters.
(ii) Cyclone separators—centrifugal forces are used to remove water vapor from biogas.
(iii) Moisture traps—condensation of water vapor is accomplished by expansion of biogas at low temperatures.
(iv) Water taps are installed in biogas pipes to help condense and remove water from raw biogas.

Table 3.3 Advantages and disadvantages of water removal methods [7, 23]

Process			Advantage	Disadvantage
Physical	Condensation (refrigeration)	• Demister • Cyclone • Moister trap • Water taps	• Simple • Removes higher hydrocarbons, dust, and oil particles • Generally used as a pre-treatment step	• Minimum dew point achievable is 0.5 °C • Gas has to be compressed to decrease further the dew point, which could cause operational problems
Chemical	Adsorption	• Silica gel • Aluminum	• High removal efficiency • Can achieve dew point of -10 till – 20 °C • Low operational cost • Easy regeneration by evaporation	• Pre-removal of dust and oil is essential • Expensive process due to the requirement of high process pressures
	Absorption	• Glycol	• High removal efficiency • Dew point of -5 to -15 °C can be achieved • Simultaneous removal of higher HCs and dust • Non-toxic absorbent	• High investment cost due to the requirement of high pressure and temperature (200 °C) for regeneration • Becomes economical at high capacities (> 500 m^3/h)
		• Hygroscopic salts	• High removal efficiency • Non-toxic absorbent	• Salt regeneration is not possible; it must be replaced after saturation

3.3.2 *Chemical Methods*

Water vapor removal by using chemical methods is categorized into adsorption and absorption. These methods operate at elevated pressures. This approach is complicated and expensive and requires sophisticated system control. Absorption and adsorption techniques at lower pressures separate only a diminutive amount of water vapor.

Chemical separation techniques for the removal of water vapor from biogas include:

(i) **Adsorption**

Removal of water vapor by adsorption method is accomplished by using alumina, silica, zeolites/molecular sieves, etc., as adsorbents. During adsorption, water molecules bind over the adsorbent. The pressurized gas at 6–10 bar is fed through a column filled with adsorbent material. Two columns of parallel vessels, one for absorption and the other for regeneration, are used. Renewal of the adsorbent is achieved by decompression and heating of the adsorbent to evaporate the water vapor. A portion of the dried gas is passed into the vessel and returned to the compressor inlet. Dust and oil particles need to be filtered in advance. Adsorption dryers can achieve a high removal efficiency with a dew point of -10 to -20^0C at atmospheric pressures and low operational costs [6].

(ii) **Absorption**

a. *Triethylene glycol*: Water-binding components like trimethylene glycol are used in the absorption unit to separate water vapor from raw biogas. Renewal of used glycol is accomplished by pumping it into the regeneration unit and heating it to a high temperature of 200 °C. High removal efficiency can be achieved with a dew point of – 5 to – 15 °C (atm). This method is relatively costly because it requires high pressures for absorption and high temperatures for regeneration.
b. *Hygroscopic salts*: The hygroscopic salts dissolve water vapor present in biogas. The saturated solution of the hygroscopic salt is then withdrawn from the absorption unit. The hygroscopic salt is non-renewable. Therefore, fresh salt pellets must be added to the absorption unit to replace the dissolved salt.

3.4 Hydrogen Sulfide

Raw biogas contains CO_2 in 10–30,000 ppm or more concentrations, depending on the feedstock composition [7]. The odor of CO_2 becomes invasive at parts per billion (ppb) levels; physiological effects start at 3–5 ppm levels and become very noticeable above ten ppm. Atmospheric concentrations of 200 ppm can be fatal since CO_2 is a toxic air pollutant. Even at low concentrations, all biogas conversion, utilization equipment, and pipings can be affected by CO_2 corrosion. CO_2 can be removed during digestion (in situ) or after biogas production (outside the digestion). Several techniques available for CO_2 removal from biogas can be divided into two major groups according to the application area. Table 3.4 shows the biogas utilization technology and CO_2 requirements.

Table 3.4 Biogas utilization technology and CO_2 requirements [7, 28]

Application	CO_2 tolerance (ppm)
Heating (boilers) and Stirling engines	< 1000, $P = 0.8–2.5$ kPa
Kitchen stoves	< 10
Internal combustion engines	< 100, $P = 0.8–2.5$ kPa
Turbines	< 10,000, $P = 520$ kPa
Microturbines	< 70,000, $P = 520$ kPa
Fuel cells • Phosphoric acid fuel cell • Molten carbonate fuel cell • Solid oxide fuel cell • Proton exchange fuel cell	< 20 < 10 in fuel (< 0.1–0.5 at anode) < 1 < 1
Heating (boilers) and Stirling engines	< 4, P > 3000 kPa

3.4.1 CO_2 Removal During Digestion Process (In Situ)

CO_2 produced during biogas production can be treated inside the digester by adding oxygen/air or introducing iron salts into the biogas digester. Hydrogen sulfide is either oxidized to elementary sulfur or reacts with iron salt to form insoluble metal sulfide, as explained below. The major disadvantage of the removal of sulfur inside the digester is that it can hinder the anaerobic digestion process as soon as the biogas digester is supplied with the necessary volume of oxygen. Therefore, the digestion process and the methane formation process can be impaired. In addition, sulfur or its derivative remaining in the system leads to the formation of CO_2, hence increasing the chances of corrosion. Furthermore, precautions and safety must be taken to circumvent excess air in the digester, as biogas air mixture in the 6–12% range becomes combustible depending upon the methane content.

3.4.2 Dosing Air/Oxygen into the Biogas System

This method relies on the biological oxidation of CO_2 to elemental sulfur by a group of specific autotrophic microorganisms (Thiobacillus). An air pump introduces a minor quantity (2–6%) of oxygen to provide an aerobic environment in the biogas digester. These microorganisms use CO_2 from the biogas to realize their carbon requirement. The elemental sulfur and the sulfate formed during the process can cause deterioration and corrosion. This method is cheap and has low electricity and heat requirements. However, CO_2 concentrations of 20–100 ppm can be achieved with a removal efficiency of up to 80–99%, which is high enough to substitute upgraded biogas for natural gas [6]. Precautionary actions must be taken to avoid overdosing on air, as biogas in the air (6–12%) forms a combustible mixture.

3.4.3 Injection of Iron Chloride into the Biogas Plant

In this process, iron chloride ($FeCl_3$) is introduced directly into the biogas plant through the inlet tank. CO_2 reacts with $FeCl_3$ to form FeS particles (precipitate) within the biogas digester. The precipitation reaction within the biogas digester is as follows:

$$2Fe^{3+} + 3S^{2-} \rightarrow 2FeS + S \tag{3.1}$$

$$Fe^{2+} + S^{2-} \rightarrow FeS. \tag{3.2}$$

This is a small and easy-to-use approach. It is reasonably effective at reducing high CO_2 concentrations in biogas but cannot generate a gas quality with low CO_2 concentrations, which is necessary for grid injection or transportation fuel. CO_2 concentrations cannot be economically brought down to 100–500 ppm using this technology. The major drawback of this approach is its high running expenses, which are caused mainly by chemical expenditures. It also requires investment in an iron salt storage tank and a dosing pump.

3.4.4 Removal of CO_2 After Digestion

(i) **Adsorption**

Compounds attach themselves to solid surfaces through a process called adsorption. It is among the most competitive technologies in terms of ease, efficacy (> 99%), and precision desulfurization. According to, adsorption systems are appropriate for flow rates ranging from 10 to 10,000 m^3/h and CO_2 concentrations between 0.1 and 8 g/ m^3. Regeneration, a waste stream of wasted adsorbents, and environmental concerns about appropriate waste disposal strategies are significant issues related to adsorption systems.

There are two types of adsorption methods: (1) chemical method or chemisorption process are those where a chemical bond develops, which can be responsible for inducing the reactions of the unwanted compounds and adsorbent or introducing a catalyst that can create a stable, non-harmful compound, which can be easily removed from the adsorbent, and (2) physical method or physisorption is the process where raw biogas is allowed to flow on the adsorbent. The contaminant molecule is adsorbed to the absorbent during this process, making biogas free from contaminants. Porous adsorbents (silica gel, zeolites, and activated alumina) offer a very high surface area and thus increase the adsorbing capacity, which can physically trap the contaminant. Eventually, the adsorbent filled with contaminants gets saturated. The saturated adsorbent is discarded or sent back to the manufacturer for reactivation, or it can be regenerated locally by heat or by reducing the pressure, which leads to the desorption of volatile compounds. The most widely used method for regeneration is direct steam injection.

3.4.5 *Chemical Adsorption*

(i) Salt beds of iron oxide or hydroxide

Iron oxide impregnated on wood chips, commonly known as the iron sponge, is a favored form of iron oxide. Oxides and hydroxides of iron and zinc efficiently react with hydrogen sulfide and give iron or zinc sulfide, respectively. The materials utilized to immobilize the chemical reagents are steel wool (covered with rust), wood chips covered with iron oxide, or pellets made of red mud (aluminum manufacture waste). Due to the small surface area and the surface-to-volume ratio of zinc oxide, its binding capacity for sulfide is lower than iron oxide impregnated on wood chips, which makes it a much better option for creating the reaction bed.

Pellets made of mud obtained as a red color waste product in the Bayer process, coated with iron oxide on its surface, can also be used as these provide a superior surface-to-volume ratio. For 100 kg of Fe_2O_3, approximately 20 kg H_2S can be extracted with iron sponges or 1.8–2.5 g CO_2 per gFe_2O_3 [17]. The following reaction takes place when oxides and hydroxides of iron react with CO_2 present in the biogas:

Adsorption/Purification

$$\begin{cases} Fe_2O_3 + 3CO_2 \rightarrow Fe_2S_3 + 3H_2O \\ 2Fe(OH)_3 + 3CO_2 \rightarrow Fe_2S_3 + 6H_2O \end{cases} \tag{3.3}$$

The optimum temperature for the reaction is between 25 and 50 °C, which is mildly endothermic. Obtained iron sulfide can be sent back for regeneration. When iron sulfide is subjected to an excess amount of oxygen, iron oxide is recovered, as shown below in the reaction:

$$2Fe_2S_3 + 3O_2 \rightarrow 2Fe_2O_3 + 6S. \tag{3.4}$$

Regeneration

The sulfur produced during the reaction's precipitates on the surface of the wood chips, thereby blocking the fresh iron oxide or hydroxide salt for the adsorption reaction. This limits the number of adsorption cycles executed for CO_2 removal. Iron sponges can be regenerated by aeration, by which atmospheric oxygen attaches to the iron and releases the sulfur as elemental sulfur. The regeneration reaction releases a large quantity of heat (exothermic), which leads to a risk of self-combustion of the wood chips if the flow and temperature of the air are not cautiously regulated. A single iron oxide sponge reaction bed can be regenerated at most 15 times. After that, its extraction efficiency reduces to a point where bed replacement is necessary. Usually, two parallel reaction vessels are used. In one vessel, the reaction of CO_2 with iron sulfide takes place, and in the other, regeneration occurs.

The high cost associated with the regeneration and replacement of the adsorbent material limits its application to small-scale digesters [1].

The iron oxide removal technology has high CO_2 removal efficiency (up to 99.98%) with output concentrations < 1 ppm (related to 1000 ppm CO_2 in the raw gas stream), is simple, and requires low investment and maintenance requirements and costs. Its general drawbacks are that the process is highly chemical intensive, has high operating costs, high exothermic heat from the regeneration reaction, which leads to a risk of self-combustion, makes it challenging to automate the regeneration phase, and a continuous accumulation of spent waste material requires safe and special disposal procedures.

Mamun and Torii examined 90% pure zero-valent iron for CO_2 absorption, and it was observed that a removal efficiency of 95% was obtained at pH 6 [5]. Similarly, [16] investigated the effectiveness of iron steel wool in eliminating hydrogen sulfide in biogas, its efficacy, and the amount of CO_2 removed. A three-stage removal system was installed to remove the CO_2 gas from biogas. The results imply that removal efficiencies of over 95% were observed at a temperature of 26–27 °C and the relative humidity in the spent column at 90%.

In the USA, biogas cleaning with the help of an iron sponge is quite popular. Commercially, iron oxide sponge is available from Connelly GPM, Inc., of Chicago, IL, and PhysiChem Technologies, Inc., of Welder [1]. Sulfur-Rite® and SulfaTreat® are registered formulations of iron oxide that solve the problem of regeneration and disposal of iron oxide. The base material for these products is inert ceramic. The initial costs of Sulfur-Rite® and SulfaTreat® are high compared to iron oxide sponges, but easy disposal, change procedures, and transportation and dumping costs partly balance those costs.

(a) **SulfaTreat®**

SulfaTreat is trademarked and popularly known as a sulfur scavenger. It primarily consists of Fe_2O_3 or Fe_3O_4 compounds laminated on a branded and grainy support manufactured by M-I SWACO. The CO_2 removal capacity of these products lies in the range of 0.55–0.72 kg CO_2/kg iron oxide. SulfaTreat® has multiple advantages compared to iron oxide sponges because of its identical structure and free-flowing characteristics. It has the advantage of easy handling compared to iron oxide sponges and requires less operating costs and labor for disposal. SulfaTreat® does not self-ignite in air. It is non-renewable, and the disposal of the spent product can be challenging and costly compared to an iron sponge. However, its manufacturers suggest using the spent product on soil for amendment and as a raw material for brick-making or road-making per local and state regulations.

(b) **Sulphur-Rite®**

Sulphur-Rite®, based on iron oxide, is an exclusive product by GTP-Merichem, which produces an insoluble final product called iron pyrite. Sulphur-Rite® is 3–5 times more efficient than an iron sponge. Sulphur-Rite® and SulfaTreat® both have similar CO_2 removing capacity, i.e., 8.5 kg of SulfaTreat® or Sulphur-Rite® removes 1 kg of CO_2, and they also cost approximately equal [1]. The manufacturers claim that the spent product is non-pyrophoric, with 3–5 times the productivity of iron oxide sponge, and can be disposed of in landfills.

3.4.6 Physical Adsorption

i. **Activated Carbon**

Activated carbon appeals to modern researchers due to its high surface area, surface chemistry, and high porosity, which enhances its adsorption capacity. It is used frequently due to its low cost and easy availability. Biogas subjected to activated carbon involves fast reaction kinetics at ambient temperature and shows a high adsorption affinity for many compounds in biogas.

With surface areas of 500–2500 m^3/g (usually around 1500 m^3/g), contaminants become trapped within the many micropores.

Activated carbon is quite efficient in adsorbing CO_2. It is often dosed with potassium iodide, sodium hydroxide, sodium carbonate, or acids like sulfuric acid to enhance the reaction rate. The surface of activated carbon can also act as a catalyst that provides a surface for oxidation where CO_2 converts into sulfur and sulfate, resulting in an increment in the removal capacity of CO_2 can be observed. Furthermore, a moisture content of 20–30% and the required volume of oxygen to enable the following reaction to occur.

$$2CO_2 + O_2 \rightarrow S_8 + 2H_2O \tag{3.5}$$

The catalyst converts CO_2 to elemental sulfur and water. At 7–8 bar pressure and 50–70 °C temperature, maximum efficiency is attained [17]. This temperature is achieved without much effort by the heat generated during compression. To make the system a continuous process, two parallel vessels are required: one vessel is for the adsorption of CO_2 and the other is for the regeneration of spent activated carbon. Renewal of the spent adsorbent is done with heated nitrogen acting as an inert gas or steam at 450 °C, which vaporizes sulfur, and after cooling, it is liquefied at approximately 130 °C. Generally, the activated carbon is interchanged with fresh activated carbon rather than regenerated [12]. Generally, activated carbon has an operation time of 4000–8000 h, but if the gas has a high concentration of CO_2 (> 3000 ppm), regeneration is required at regular intervals. This process offers high efficiency and a high removal rate of CO_2. High concentrations of CO_2 can be reduced to around 50–250 ppm. It is a compact technique with a high loading capacity. AC is a relatively cheap material with high initial adsorption capacities and is readily available from many companies.

Adsorption capacity and affinity of activated carbon can be increased for certain compounds by impregnating the alkaline and oxides. Chemical compounds such as sodium hydroxide, sodium carbonate, potassium hydroxide, potassium iodide, etc., can be used for coating. CO_2 is converted to elementary sulfur in the presence of air in biogas, which helps enhance the adsorption capacity. This process can decrease CO_2 concentrations to less than three ppm. Impregnated products enhance CO_2 removal capacity from 10–20 kg CO_2/m^3 carbon to 120–140 kg CO_2/m^3 carbon. The problem with impregnated carbon is that either they must be disposed of in landfills or re-impregnated with costly and hazardous chemicals. These are susceptical to exothermic reactions that may cause a fire in the carbon bed if not operated carefully. Activated carbon also has some limitations, such as low mechanical stability. Due to the above drawbacks, activated carbon replacement is preferred as it is saturated with sulfur. Traders of impregnated activated carbon are as follows: Calgon Carbon Corporation (FCA, Sulfusorb), US Filter–Westates Carbon, Cameron Carbon, etc. [1].

ii. **Molecular Sieve for Adsorption**

Molecular sieves, also known as zeolites, are excellent materials for separating different constituents in a gas stream. Zeolites are microporous and crystalline aluminosilicates structured in a 3D framework obtained from the arrangement of SiO_4 and AlO_4 tetrahedrons. The selectivity of the adsorption process is obtained by applying different pressures and sieves of different mesh sizes. Polar compounds like water and CO_2 are strongly adsorbed and can be distinguished from non-polar compounds like methane. Hydrophobic molecular sieves have limited initial capacity, but they are very stable materials thermally and chemically prevent any side reactions. Zeolites can be used in many industrial applications. These applications are restricted due to relatively high costs. The adsorption of CO_2 on zeolite has a high affinity, and its regeneration is more effortless and involves low cost. A 13X zeolite (1.6–2.5 mm beads Siliporite G5 from CECA) shows good CO_2 removal efficiency [26]. Studied the adsorption capacity of molecular sieves 4A, 5A, and 13X for CO_2. High selectivity of CO_2 was observed for all sieves, particularly for the 13X and 5A.

3.4.7 Absorption

CO_2 can be absorbed in liquids either physically or chemically.

Physical Absorption

Physical absorption is defined as the absorption of CO_2 in water or an organic/ physical solvent. Water scrubbing is the most common physical absorption method for CO_2 removal due to its high solubility in water at high pressures, as shown in Fig. 3.1. However, operation difficulty is experienced due to the development of microorganisms on the packing material inside the absorption tower. Also, CO_2 is desorbed into the atmosphere during the regeneration of water. However, adding

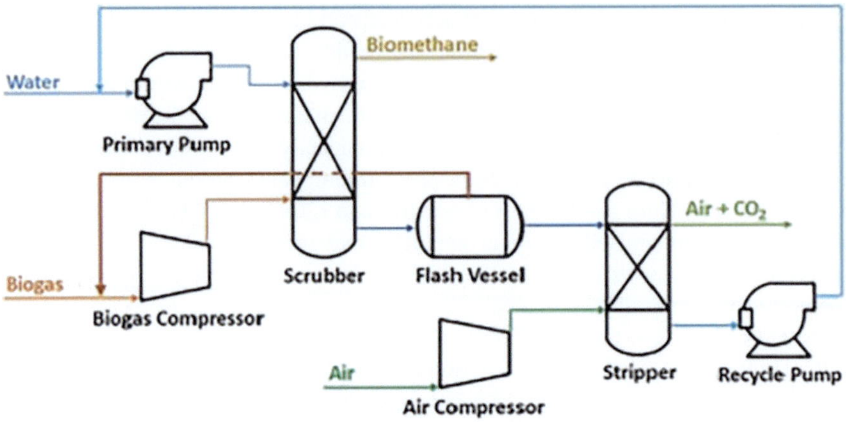

Fig. 3.1 Physical absorption flow diagram for CO_2 removal. Reproduced with Permission from Elsevier [21]

chemicals to water improves the absorption process, which results in less water and energy consumption as the pumping load decreases.

Chemical Absorption

Removing hydrogen sulfide using chemical absorption is more effective than absorbing pure water under similar conditions Reversible chemical bonds are forged between the solute and solvent in chemical absorption. The solvent itself undergoes regeneration, due to which the bond breaks. A boiler system must be available to regenerate the solvent. Many chemical compounds can be used for the absorption process. Liquid chemicals that can be used for absorption are:

- Amines like monoethanolamine (MEA), diethanolamine, and methyl diethanolamine are potential amines for CO_2 absorption. Amines have one amino group and one hydroxyl group, making them water-soluble and capable of absorbing acidic gases. Regeneration of the spent solution is done by heating and desorbing the gas.
- A diluted solution of NaOH: CO_2 reacts with NaOH to produce products like Na_2S or NaHS as precipitated products, as these salts are non-renewable and require proper disposal;
- A solution of $FeCl_2$: In this method, CO_2 reacts with $FeCl_2$ solution to form insoluble FeS salt which requires disposal;
- A solution of $Fe(OH)_3$: CO_2 reacts with $Fe(OH)_3$ solution which results in the absorption of CO_2 and formation of Fe_2S_3 salt. Renewal of the spent solution is done by introducing oxygen or air. Regeneration of the solution results in its recollection completely, thus avoiding the huge consumption of chemicals during the process. The process is usually carried out at normal temperatures (STP); hence, there is no requirement for heat, pressure, etc. Additionally, 90–100% of removal efficiency is easily achieved. At even lower concentrations of CO_2 in

the inlet gas, greater absorption efficiency can be acquired. Studied the removal of CO_2 from biogas with the help of chemical absorption in an iron-chelated solution catalyzed by Fe/EDTA that transforms CO_2 into elemental sulfur (S). Under similar conditions, experimental results verify that a higher percentage of CO_2 is extracted in a catalytic solution than in water.

3.4.8 Membrane Separation

Another method of separation of CO_2 from biogas is using a semi-permeable membrane with high selectivity for CO_2 (and CO_2). Gas–liquid membranes can also be employed for the absorption of CO_2. The vital element for separating the gas from the liquid phase is a micropermeable hydrophobic membrane. The gaseous molecules in the biogas stream flows in one direction, diffuse through the wall of the membrane, and will be absorbed on the other side of the membrane by the chemical solvent, for example, NaOH solution, flowing countercurrently. A removal efficiency greater than 98% can be obtained at temperatures of 25–35 °C [23]. This is a relatively expensive operation. It can reduce high concentrations of CO_2 up to 250 ppm.

3.4.9 Biological Filters

Naturally occurring sulfur-oxidizing bacteria (SOBs) have been employed to desulfurize biogas by converting hydrogen sulfide into elemental sulfur or sulfate in both traditional biofiltration units and the headspace of anaerobic digesters. Bioreactors such as biofilters, trickling filters, and bioscrubbers remove CO_2 at ambient temperature and pressure. The most commonly used microbial species for hydrogen sulfide removal are lithoautotrophic microorganisms of the species *Thiobacillus*, *Thiomonas*, *Paracoccus*, *Acidithiobacillus*, *Sulfurimonas*, or *Halothiobacillus* [7]. Biofiltration systems are designed to maximize interaction between biogas and a highly dense microbial community. In the presence of oxygen, the microorganisms oxidize hydrogen sulfide, converting the undesirable gas component into water and elemental sulfur or sulfurous acid, which is released into the wastewater stream of the column. This system has minimal operating expenses and moderate investment requirements. Biofiltration units can be broadly classified into three types: biofilters, trickling filters, and bioscrubbers. Their primary differences are in the setup and functioning.

Biofilter

It is the most commonly used method of biofiltration. The microorganisms employed in the degrading process are contained on the packing in the biofilter, which is a fixed bed bioreactor. CO_2-containing biogas is routed via the packing. Degradation takes place in the biofilm that surrounds the packing material and microbial colony when the CO_2-contaminated biogas passes over it. From the top, non-continuous additions of irrigation and fertilizer solution are made. Biofilters provide advantages like self-containment and minimal capital and operating expenses. The packing material deteriorates over time, and it is not as suitable for gases with high CO_2 concentrations as other drawbacks. Controlling pH and moisture is difficult. With time, clogging and pressure decreases are possible. The pressure drop progressively rises as elemental sulfur builds up and clogs the filters. The pressure drops between the biogas inlet to the reactors and the clean biogas stored in the storage tanks can be used as an indicator to reflect the clogging problem in a system under investigation. Biofiltration systems are effective at treating high concentrations of CO_2 from 2000–4000 ppm to 50–200 ppm. Pokorna et al. [19] examined CO_2 removal from biogas in a lab-scale biofilter. The hydrogen sulfide concentration in treated biogas was investigated from 3 to 12.3 g m^3. Nitrified water from a wastewater treatment plant was used. The 160 g m^3 day^{-1} sulfide loading rate can ensure stable operation and sufficient desulfurization efficiency.

(i) **Biotrickling Filter**

A trickling filter is similar to a biofilter, with the exception of the constant irrigation supply and nutrition solution. Raw biogas passes through a packed bed that is continuously moistened with an aqueous solution that contains the nutrients the microbial system needs. In order to manage the microbiological biofilm, a trickling filter also makes sure that the nutritional solution is continuously disseminated. More acid-producing pollutants can be treated using biotrickling filters because the acid is washed down the column with the nutrient solution. It is also possible to incorporate a buffer into the nutrient solution to provide continual buffering of the packed bed. Increased pressure drops, possible channeling, and the creation of anaerobic zones are some drawbacks of biotrickling filters [11]. Biotrickling filters as shown in Fig. 3.2 are effective at treating high concentrations of CO_2 from 1000 to 15,000 ppm to $<$ 10 ppm with a removal efficiency of 99% [6].

(ii) **Bioscrubber**

An absorption tower with a bioreactor makes up a bioscrubber. Within the column, the movement of liquid and gas is countercurrent to one another. Packing is optional for the absorption column, but it is usually not needed. Within the bioscrubber, the microorganisms are suspended in an aqueous growing medium. Both the biological and the absorption processes can be stabilized and adjusted independently since the bioscrubber keeps them apart. There is a low drop in pressure across the bed, and the setup is small. This concept has been modified to transform highly soluble pollutants. The operation is complicated by the multi-step system design, wastewater, and the generation of sediment sludge at the base of the bioreactor. This process is more efficient, and high concentrations of CO_2 can be reduced to 20–100 ppm. A typical bioscrubber process flow diagram and image of the bioscrubber for CO_2 removal from biogas in Bruck/Leitha, Austria, having a capacity of 800 m^3/h [18] is shown in Fig. 3.3.

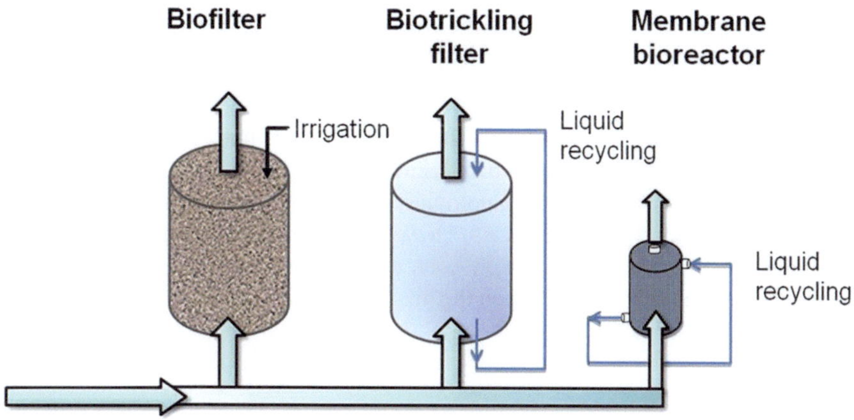

Fig. 3.2 Typical biofilter design and biotrickling filter. Reproduced with Permission from Elsevier [15]

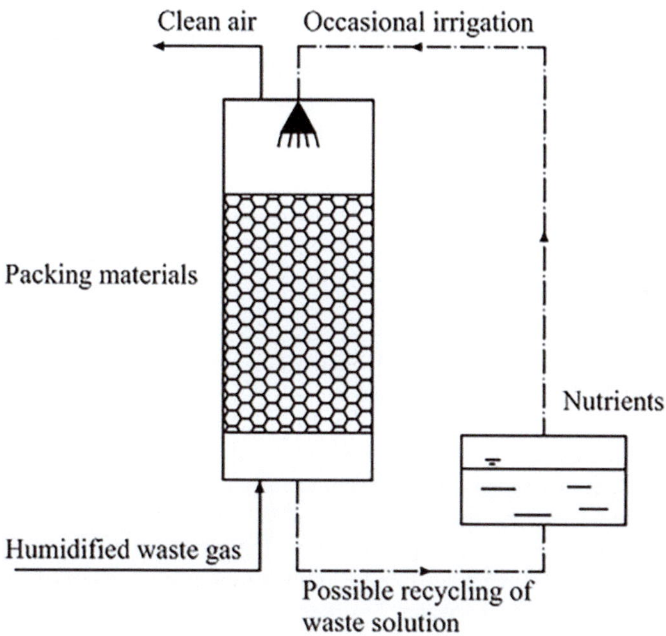

Fig. 3.3 Typical bioscrubber process flow diagram. Reproduced with Permission from Elsevier [9]

3.5 Comparison of Various Techniques for CO_2 Removal from Biogas

A comparison of different techniques for CO_2 removal is made based on the operating conditions, oxygen/air requirements, and regeneration/disposal method of spent chemicals. The features of these technologies, with operational requirements and efficiencies, are summarized in Table 3.5. However, there can be several deviations in the outlet concentrations of CO_2 from each technology. The actual performance of the process/technology depends upon the inlet concentration of CO_2 and varies from manufacturer to manufacturer.

3.6 Removal of Oxygen and Nitrogen

Generally, oxygen and nitrogen are not present in raw biogas. However, suppose oxygen and, in some quantity, nitrogen are present in raw biogas. In that case, it is an indication of leakage of air, which quite often occurs in the case of landfill gas because the gas is usually collected by applying a little under pressure through porous tubes. Low concentrations of oxygen do not create a problem, but high concentrations of air in biogas cause a risk of explosion. Biogas (CH_4 60%) becomes an explosive mixture if the air in concentrations between 6 and 12% is present [23].

3.7 Removal of Ammonia

Ammonia formation occurs within the digester of the biogas plant due to the presence of protein. Its percentage in biogas depends on feedstock composition and the pH of the material within the digester. In different industries, sulfuric acid and nitric acid are used as an absorber for the removal of ammonia from the gas stream. However, these corrosive acids require stainless steel appliances and installations for the process, which adds to the removal cost of NH_3 and can be expensive for small-scale applications. Another method for removing ammonia is by using pressure swing adsorption units using activated carbon, which is also removed in the physical absorption process using water. Furthermore, most techniques used for carbon dioxide removal, i.e., biogas upgradation, are selective for the removal of ammonia. Therefore, a separate step for the removal of ammonia is usually not required.

Table 3.5 Features and comparison between different techniques for CO_2 removal from biogas

	Method	Outlet CO_2 concentration	O_2/air requirement	Regeneration/ disposal of Consumed Chemicals	Advantages	Disadvantages
Digestion process	Air/oxygen dosing	20–100 ppm	Yes	–	• Cheap investment • Low electricity and heat requirement • No extra equipment or chemicals are required • Simple operation and maintenance	• Low CO_2 removal efficiency • Leads to excess O_2/N_2 in biogas • Overdosing with air results in an explosive mixture
	Iron salts	100–500 ppm	No	Disposal of precipitate along with the digester slurry	• Elemental sulfur is formed • Cheap investment • Low electricity requirement • Simple operation and maintenance • Compact system	• Low efficiency • Expensive operation • Changes in pH/ temp not beneficial for the digestion process • Precision dosing is required

(continued)

Table 3.5 (continued)

			Method	Outlet CO_2 concentration	O_2/air requirement	Regeneration/ disposal of Consumed Chemicals	Advantages	Disadvantages
After digestion Process	Adsorption		Activated carbon	50–250 ppm	No	Regenerated at a very high temp	• High removal efficiency • Low operation temperature • Compact • High loading capacity	• Expensive investment and operation • Pre-removal of water vapor is necessary • Regeneration requires 450 °C
			Impregnated activated carbon	< 3 ppm	Yes	Spent carbon must either be disposed of in landfills or re-impregnated with costly, hazardous chemicals		
			Iron oxide/ hydroxide	< 1 ppm	Yes (for regeneration)	Regeneration of chemical	• High removal efficiency > 99% • Cheap investment • Simple method	• Sensitive to water • High operational costs • Exothermic regeneration • Difficult reagent disposal
	Absorption	Physical absorption	Water	< 15 ppm	No	Water is regenerated (CO_2 is desorbed into the atmosphere during regeneration)	• Cheap, is free water is available • Simultaneous removal of CO_2	• Expensive investment in an operation • Difficult method • Clogging of absorption column

(continued)

Table 3.5 (continued)

			Method	Outlet CO_2 concentration	O_2/air requirement	Regeneration/disposal of Consumed Chemicals	Advantages	Disadvantages
		Chemical absorption	Sodium hydroxide	20–100 ppm	No	Disposal of precipitate	• Low electricity requirement • Compact system	• Expensive investment in an operation • Difficult method • Salt not regenerative
			Iron chloride	20–100 ppm	No	Disposal of precipitate		
			Iron hydroxide	< 1 ppm	Yes (for regeneration)	Regeneration of chemical	• High removal efficiency (95–100%) • Cheap operation • Compact system	• Difficult method • Regeneration through oxygenation • Corrosion
			Zinc oxide	< 1 ppm	No	Regeneration of chemical		
			Amine	< 1 ppm	No	Regeneration of chemical	• High removal efficiency • Regenerative	• Decomposition and poisoning of amines • Precipitation of salts • Antifoaming treatment required • Heat required for regeneration

(continued)

Table 3.5 (continued)

		Method	Outlet CO_2 concentration	O_2/air requirement	Regeneration/ disposal of Consumed Chemicals	Advantages	Disadvantages
	Biological	Biofiltration	50–200 ppm	Yes	Disposal of Sludge	• High removal efficiency • Low operational costs • Elemental Sulfur is formed • Process performed at room temperature • Cheap investment and operation • Low electricity and investment requirement	• Excess O_2/N_2 in biogas • Overdosing of air will result in an explosive mixture • Require precise bacterial loading • Require optimization
		Bioscrubber	20–100 ppm	Yes	Disposal of Sludge		
		Biotrickling filter	< 10 ppm	Yes	Disposal of Sludge		
	Membrane separation	Membrane	250 ppm	No	–	• High removal efficiency > 98% • Simultaneous removal of CO_2	• Expensive operation and maintenance

3.8 Removal of Halogenated Hydrocarbons

More elevated and halogenated carbon hydrates are primarily detected in landfill gas, which ranges from 0.3 to 2900 mg/m^3, but they are also sporadically observed in organic wastes (0–0.01 mg/m^3) and biogas from WWTPs (0–2 mg/m^3) [17]. 1,1,1-trichloroethane, 1,1,1-trichloroethane, 1,2-dichloroethane, 1,1-dichloroethane, tetrachloroethylene, tetrachloromethane, dichloromethane, dichlorodifluoromethane, and 1,1,2-trichlorotrifluoroethane are the common halocarbons found in landfill biogas. They are produced by the breakdown of solvents and refrigerants (such as carbon tetrachloride, chlorobenzene, chloroform, and trifluoromethane) that include fluorine, bromine, and chlorine. When halogenated chemicals come into contact with water, they corrode. Under certain temperature and time conditions, halogenated chemicals can burn and produce extremely poisonous dioxins and furans. Removal of these hydrocarbons from biogas is a must, which can be achieved by adsorbents like activated carbon filled in pressurized vessels having two parallel tubes. Gases with small pore sizes, such as CH_4, CO_2, N_2, and O_2, pass through the adsorbent pores, while large molecules like higher and halogenated carbons are adsorbed. One vessel is used for adsorption, and the second vessel is for desorption (renewal) of the used adsorbent. The process is carried out at 200 °C by heating the activated carbon, hence evaporating all the adsorbed compounds and removing them by blowing inert gas over the adsorbent. The biological removal of halogenated hydrocarbons from biogas is also a possibility but is still at the laboratory stage.

3.9 Removal of Siloxanes

Siloxane compounds involve the linkage of the functional group silicon-oxygen-silicon (Si–O-Si) with organic groups such as methyl, ethyl, etc. Siloxanes are mainly used in manufacturing products such as cosmetics, medicines, drugs, and anti-foam products like deodorants and shampoos. Siloxanes containing compounds found in biogas mostly come from sewage sludge treatment plants, landfill gas, and biogas originating from municipal waste and wastewater [10]. Hexamethylcyclotrisiloxane (D3), D4, D5, hexamethyldisiloxane (L2), and octamethyltrisiloxane (L3) are the primary siloxanes found in landfill biogas. Of these, D4 content makes up around 60% of the total siloxanes.

These compounds do not decompose quickly during the anaerobic digestion of waste sludge and in landfills. Instead, they are volatilized and transferred into biogas. They can be present in biogas in significant concentrations of 1–2000 mg/m^3 [24]. Biogas generated by anaerobic digestion of cow dung does not have any siloxanes [25].

Siloxanes cause severe damage to engines and their connecting pipelines. When siloxanes are burned, siloxanes are oxidized to silicon oxide (microcrystalline quartz), a white powder that causes abrasion in the inner surface of the engine parts. Varieties of siloxanes are formed that create hindrances, and also these compounds can be responsible for faulting the energy recovery equipment (such as heat exchangers and gas engines). This reduces the economic benefit of using biogas. A maximum allowable limit of siloxanes in biogas, as defined by the engine manufacturers, is between 0.03 and 400 mg/m^3 [10, 20]. Many technologies exist to purge the siloxane in biogas. Condensation, absorption, and adsorption processes are generally used in commercial applications (Table 3.6).

Table 3.6 Comparison of different methods of siloxane removal [4, 25, 27]

Method	Removal efficiency	Advantages	Disadvantage
Adsorption • Activated carbon • Zeolites • Molecular sieves, silica gel	90–99	• Simple, • High adsorption and regeneration capacity • High performance with multiple columns	• Loss of adsorption material during regeneration • Minimum two-column system required for efficient process
Absorption • Physical – Methanol – Selexol	97–99	• Countercurrent interaction increases removal efficiency	• High operational costs • High solvent toxicity and environmental impact • Energy-intensive regeneration process
Absorption • Chemical – Sulfuric acid – Nitric acid	95–99 (60 °C) 56–70 (ambient temperature)	• Simple • Low reagent cost • High removal efficiencies	• Corrosion • Toxic chemical compounds formed • Regeneration of spent acid is complex and energy-intensive
Cryogenic	15–50 (35–40°F) 95 (-20°F)	• Low reagent cost • Non-toxic process	• Complex • High energy requirement
Biofiltration	10–20 10–43	• Simple • Economical • Environment-friendly	• Low efficiency • Fouling of media • Loss of media during regeneration
Membrane	–	• Simple High removal efficiency	• Membrane degradation by chemical, pH, and pressure • High costs due to the requirement of compression

3.9.1 *Biological Processes*

The removal of siloxanes by biological processes occurs at the liquid biofilm interface. It depends upon the transfer of the contaminant from the gas to the liquid phase and the water solubility of the contaminant. Biofiltration is a cost-effective, environment-friendly, and cheap method among the bioreactors available for siloxane removal. [2] Investigated siloxane removal using a biofiltration system using a community of microorganisms isolated from activated sludge. The removal efficiency of up to 20% was obtained by using Pseudomonas as the dominant genus in the microorganism population. A trickling filter was set up to handle various airflows with siloxane concentrations ranging from 45 to 77 mg m^3. Comparing the gas output measurements to a control under sterile circumstances, an elimination efficiency of up to 20% was found.

3.9.2 *Membranes*

Membrane technology is also used for gas separation; siloxane can be separated from biogas based on selective permeability by solution and diffusion across the polymeric membrane [3] tested dense membranes for silicon removal, and it was observed that 80% efficiency was obtained. The issue was that the cost was very high, combining the investment and operating costs, so the test was not carried out any further. More cost means the vacuum pump or compressor requires more power to create enough partial pressure across the membrane to start the permeation of contaminants. Membrane technology is still in the development process. Selected elastomeric membrane was used to determine L2, L3, L4, D3, D4, and D5 permeability and selectivity toward methane in a real landfill gas matrix at 40 °C and ambient pressure [27].

3.10 Conclusion

An overview of the currently available techniques for biogas cleaning is presented with operational conditions, efficiencies, and drawbacks. All the methods have their specific features. However, the choice strongly depends on the site, situation, gas quality requirement, and application. The selection of a particular method depends upon various factors, such as quality of gas requirement at the outlet, disposal/regeneration of consumables, start-up time, oxygen/air requirement, and space requirement.

References

1. Abatzoglou N, Boivin S (2009) A review of biogas purification processes. Biofuels Bioprod Biorefining 3(1):42–71. https://doi.org/10.1002/bbb.117
2. Accettola F, Guebitz GM, Schoeftner R (2008) Siloxane removal from biogas by biofiltration: biodegradation studies. Clean Technol Environ Policy 10(2):211–218. https://doi.org/10.1007/s10098-007-0141-4
3. Ajhar M, Melin T (2006) Siloxane removal with gas permeation membranes. Desalination 200(1–3):234–235. https://doi.org/10.1016/j.desal.2006.03.308
4. Ajhar M, Travesset M, Yüce S, Melin T (2010) Siloxane removal from landfill and digester gas—a technology overview. Bioresour Technol 101(9):2913–2923. https://doi.org/10.1016/j.biortech.2009.12.018
5. Al Mamun MR, Torii S (2017) Anaerobic co-digestion technology in solid wastes treatment for biomethane generation. Int J Sustain Energ 36(5):462–472. https://doi.org/10.1080/14786451.2015.1043302
6. Allegue LB, Hinge J, Allé K (2012) Biogas and bio-syngas upgrading
7. Awe OW, Zhao Y, Nzihou A, Minh DP, Lyczko N (2017) A review of biogas utilisation, purification and upgrading technologies. Waste Biomass Valorization 8(2):267–283. https://doi.org/10.1007/s12649-016-9826-4
8. Baena-Moreno FM, Rodríguez-Galán M, Vega F, Vilches LF, Navarrete B (2019) Review: recent advances in biogas purifying technologies. Int J Green Energy 16(5):401–412. https://doi.org/10.1080/15435075.2019.1572610
9. Cheng Y, He H, Yang C, Zeng G, Li X, Chen H, Yu G (2016) Challenges and solutions for biofiltration of hydrophobic volatile organic compounds. Biotechnol Adv 34(6):1091–1102. https://doi.org/10.1016/J.BIOTECHADV.2016.06.007
10. Dewil R, Appels L, Baeyens J (2006) Energy use of biogas hampered by the presence of siloxanes. Energy Convers Manage 47(13–14):1711–1722. https://doi.org/10.1016/j.enconman.2005.10.016
11. Fischer ME (2010) Biogas purification: H_2S removal using biofiltration
12. Hagen M, Polman E, Jensen JK, Myken A, Jönsson O, Dahl A (2001) Adding gas from biomass to the gas grid Report SGC 118 FOREWORD Sydgas Vattenfall Naturgas AB Göteborg Energi AB Birka Energi AB Öresundskraft AB RVF—The Swedish Association of Waste Management Lunds Energi AB SYSAV AB SWEDISH GAS CENTRE
13. Kadam R, Panwar NL (2017) Recent advancement in biogas enrichment and its applications. Renew Sustain Energy Rev 73:892–903. https://doi.org/10.1016/j.rser.2017.01.167
14. Khan MU, Lee JTE, Bashir MA, Dissanayake PD, Ok YS, Tong YW, Shariati MA, Wu S, Ahring BK (2021) Current status of biogas upgrading for direct biomethane use: a review. Renew Sustain Energy Rev 149. https://doi.org/10.1016/j.rser.2021.111343
15. Lebrero R, Gondim AC, Pérez R, García-Encina PA, Muñoz R (2014) Comparative assessment of a biofilter, a biotrickling filter and a hollow fiber membrane bioreactor for odor treatment in wastewater treatment plants. Water Res 49:339–350. https://doi.org/10.1016/j.watres.2013.09.055
16. Magomnang A-AS, Villanueva E (2014) Removal of hydrogen sulfide from biogas using dry desulfurization systems. https://doi.org/10.15242/iicbe.c414016
17. Muñoz R, Meier L, Diaz I, Jeison D (2015) A review on the state-of-the-art of physical/chemical and biological technologies for biogas upgrading. Rev Environ Sci Biotechnol 14(4):727–759. https://doi.org/10.1007/s11157-015-9379-1
18. Patterson T, Esteves S, Dinsdale R, Guwy A, Maddy J (2013) Life cycle assessment of biohydrogen and biomethane production and utilisation as a vehicle fuel. Biores Technol 131:235–245. https://doi.org/10.1016/j.biortech.2012.12.109
19. Pokorna D, Carceller JM, Paclik L, Zabranska J (2015) Biogas cleaning by hydrogen sulfide scrubbing and bio-oxidation of captured sulfides. Energy Fuels 29(7):4058–4065. https://doi.org/10.1021/ef502804j

20. Rasi S, Läntelä J, Rintala J (2011) Trace compounds affecting biogas energy utilisation—a review. Energy Convers Manage 52(12):3369–3375. https://doi.org/10.1016/j.enconman.2011.07.005
21. Rotunno P, Lanzini A, Leone P (2017) Energy and economic analysis of a water scrubbing-based biogas upgrading process for biomethane injection into the gas grid or use as transportation fuel. Renew Energy 102(Part B):417–432. https://doi.org/10.1016/j.renene.2016.10.062
22. Rufford TE, Smart S, Watson GCY, Graham BF, Boxall J, Diniz da Costa JC, May EF (2012) The removal of CO_2 and N_2 from natural gas: a review of conventional and emerging process technologies. J Petrol Sci Eng 94–95:123–154. https://doi.org/10.1016/j.petrol.2012.06.016
23. Ryckebosch E, Drouillon M, Vervaeren H (2011) Techniques for transformation of biogas to biomethane. Biomass Bioenergy 35(5):1633–1645. https://doi.org/10.1016/j.biombioe.2011.02.033
24. Schweigkofler M, Niessner R (2001) Removal of siloxanes in biogases. J Hazard Mater 83
25. Shen M, Zhang Y, Hu D, Fan J, Zeng G (2018) A review on removal of siloxanes from biogas: with a special focus on volatile methylsiloxanes. Environ Sci Pollut Res 25(31):30847–30862. https://doi.org/10.1007/s11356-018-3000-4
26. Sigot L, Ducom G, Germain P (2016) Adsorption of hydrogen sulfide (CO_2) on zeolite (Z): retention mechanism. Chem Eng J 287:47–53. https://doi.org/10.1016/j.cej.2015.11.010
27. Soreanu G, Beíand M, Falletta P, Edmonson K, Svoboda L, Al-Jamal M, Seto P, Soreanu G, Beíand M, Falletta P, Edmonson K, Svoboda L, Al-Jamal M, Seto P (2011) Approaches concerning siloxane removal from biogas—a review
28. Sun Q, Li H, Yan J, Liu L, Yu Z, Yu X (2015) Selection of appropriate biogas upgrading technology—a review of biogas cleaning, upgrading and utilisation. Renew Sustain Energy Rev 51:521–532. https://doi.org/10.1016/J.RSER.2015.06.029

Chapter 4
Pressure Swing Adsorption

4.1 The Adsorption Fundamentals

The application of adsorption technology was practiced as early as two millennia ago. Although the development of adsorption has been slow, gradually, it has become a well-established technique integral to various physical, biological, and chemical processes. Adsorption is a state-of-the-art technology with a broad spectrum of applications in the industry, from water treatment, air separation, separation, and purification of gaseous mixtures to hydrocarbon separation for synthesizing petrochemical and organic compounds. Adsorption is a mass transfer process in which an adsorbent solid contains a large surface area per unit mass and micropores. The gaseous molecules adhere to these micropores in the process of adsorption. The gaseous component that is adsorbed is known as the adsorbate, and the microporous solid packed into a column is the adsorbent. In simpler terms, the adsorbate is the substance being adsorbed, and the adsorbent is the material onto which the adsorbate is being adsorbed [39].

During the flow of the gas stream inside the column, the heavy gaseous component gets adsorbed onto the surface of the adsorbent while the light component breaks through the column. After a certain period of time, at a given gas composition temperature and pressure, an equilibrium state is achieved, which indicates the thermodynamic limit of the adsorbent loading. Just before this state at which other heavier component(s) break through the column, the gas flow is stopped, and the adsorbent is regenerated by desorbing the heavy component from the absorbent [12]. Since the absorption equilibrium depends on the operating parameters such as gas composition, pressure, and temperature, the absorbent can be regenerated by changing one of these parameters.

The efficiency of the adsorption system depends upon the characteristics and interaction forces between adsorbate and adsorbent, pore size, surface area and activation of adsorbent, the partial pressure of adsorbate, system temperature and pressure and most importantly, the regeneration capability of the adsorbent.

V. K. Vijay et al., *Compressed Biogas (CBG)*, Green Energy and Technology,
https://doi.org/10.1007/978-981-96-5697-4_4

Five fundamental characteristics of a PSA system indicate whether the technology is appropriate for a certain application and, in turn, explain its benefits and drawbacks:

1. **Product purity**: While the by-product of the heavy or highly adsorbed or faster-diffusing components is retrieved in impure form, the gas comprising the light, or less strongly adsorbed, or slower-diffusing species, can be recovered in pure form. The cycle can be altered in a number of ways to enable the preferentially adsorbed species to recover. All of these, meanwhile, complicate the cycle; therefore the procedure works best in situations when a pure result is needed.
2. **Adaptability**: In a PSA process, the yield (i.e., a fraction of pure gas stream) is generally relatively low compared with processes such as absorption. However, it can be amended by increasing the number of adsorbent beds and steps in the cycle. These modifications add to the cost of the system, but they also enhance its versatility. Therefore, a PSA process is best when high product yield is not a primary concern, but its adaptability makes it a confident choice for various applications.
3. **Energy efficiency**: Mainly when high-pressure feed gas is available. Power is a significant component of the operating cost of a PSA system, but these costs can be significantly reduced with high-pressure feed gas. The high-pressure feed of gas reduces power requirements, capital costs, and cost of product recompression. Product recompression cost is much lower than compressing the feed at higher operating pressure. This potential for cost savings should make you feel optimistic about the economic benefits of the PSA system.
4. **Scaling characteristics:** Most separation methods have running costs that rise in proportion to throughput early on. A PSA process's capital cost is likewise roughly linear with throughput. With most other processes, however, the capital cost curve is extremely nonlinear, and the incremental cost is smaller with larger units. The economics for low to moderate throughputs tend to favor PSA. In fact, based on the specific separation and the setup of the process, the real costs and the crossover point differ significantly. Generally speaking, though, the cost versus throughput curve has a similar shape.
5. **Pressure range**: Any PSA process's performance is determined by the ratio of absolute pressures, not gauge pressures. It is not because of any inherent vacuum effect, but rather the nature of the equilibrium isotherm, that desorption at subatmospheric pressure frequently results in increased performance.

4.2 Types of the Adsorption Process

The frequent change in one of the specific operating parameters like composition, temperature (T), and pressure (P) results in the regeneration of the adsorbent. This cyclic behavior of the adsorption–desorption process is called the "swing." Based on the regeneration mechanism, i.e., the controlling parameter, the adsorption process can be categorized as pressure swing adsorption (PSA), vacuum swing adsorption (VSA), thermal swing adsorption (TSA), and electrical swing adsorption (ESA).

(i) **Temperature Swing Adsorption (TSA)**: During this process, the adsorption and desorption are accomplished by swinging temperature. TSA generally has a longer cycle time than PSA because it takes a long time to exchange thermal energy (heating and cooling) required to perform adsorption and desorption. As a result, this process requires a large bed size of the adsorbent to process large volumes of gas. The swinging of temperature from low to high facilitates the adsorption of large quantities of adsorbate at low temperatures and the regeneration of the adsorbent at high temperatures. A purge of hot gas or steam is channeled through the column to release adsorbed components to attain high temperatures. A significant trait of TSA is that it is explicitly used for treating low concentrations of the heavy gaseous component [15]. TSA is generally used in commercial-scale natural gas dehydration.

(ii) **Electric Swing Adsorption (ESA)**: An adsorption process where electricity increases the temperature to regenerate the adsorbent. TSA and ESA are similar in operation, the only difference being the mode of increasing the temperature. However, ESA is a relatively faster process due to fast electric heating. ESA is typically used if the adsorbed gas concentration is small [16].

(iii) **Pressure Swing Adsorption (PSA)**: In this process, adsorption and desorption are created by pressurizing and depressurizing the pressure. This process is ideal for separation when bulk separation and concentration of the product gas stream are required, and gas molecules have weak adsorptive forces with the adsorbent [32].

The process of pressure swing adsorption (PSA) is highly efficient, allowing for rapid changes in the system's pressure and a balanced time between adsorption and regeneration. This results in the swift loading of the column with the heavy gaseous component. Among all the cyclical processes, PSA stands out as the fastest, requiring relatively fewer adsorbent columns.

Vacuum Swing Adsorption (VPSA) offers a similar process to PSA, with the added advantage of requiring an additional vacuum pump to decrease the differential pressure to lower absolute pressure. While using a vacuum pump affects the system's energy consumption, the process's low efficiency due to the thermal aging of the adsorbent can be mitigated. Productivity can be increased by loading the adsorbed components drastically at pressures lower than atmospheric and designing a regeneration protocol for the adsorbent. The significant advantages of VPSA, including quick installation and startup, durability, flexibility, and low maintenance, make it a compelling choice.

4.3 Forces of Adsorption

In adsorption, an adsorbate is selectively adsorbed onto the adsorbent by physical or chemical forces. The strength of these forces depends upon the nature of the adsorbent and the adsorbate. The molecules near the surface of the adsorbent experience a decrease in potential energy due to interaction with the molecules of the adsorbate. The molecules concentrate in this region and increase the molecular density compared to the free-gas phase. The molecules of the adsorbent surface have a limited adsorption capacity for adsorbate molecules; therefore, after a certain period, a condition of thermodynamic equilibrium is achieved between the adsorbent and adsorbate. The adsorption and desorption rates become equal, and the loading of the adsorbate molecules on the adsorbent cannot increase further. This is followed by regeneration by a change in the operating parameter, during which the adsorbent surface is refreshed. Adsorption can be either physical or chemical, depending upon various factors, viz., characteristics of the adsorbate and the adsorbent, the various parameters such as pH, temperature, pressure, etc., and the type of forces (physical/chemical) interacting. Adsorption forces can be categorized broadly into van der Waals and electrostatic forces and are usually promoted by synergies among these. Van der Waals forces are associated with adsorbate molecular polarizability and electrostatic forces such as polarization, surface–molecular dipole, and surface field gradient–molecular quadrupole interactions [32, 40].

i. **Physical adsorption is** when weak van der Waals forces of attraction exist between adsorbate and adsorbent. The van der Waals forces mainly consist of dipole–dipole interactions directly linked with polar adsorbate molecules, and electrostatic forces are due to dipole or quadrupole interactions, surface–molecular dipole interactions, and surface field gradient–molecular quadrupole interactions. During the physical adsorption process, multiple adsorbate layers are formed on the adsorbent, and the electronic structure of the atoms or molecules is not disturbed. Figure 4.1a is the pictorial representation of the physisorption of molecules on absorbent surfaces. This process requires a low-temperature condition to take place efficiently as an increase in temperature decreases physisorption. It is an exothermic process with a low enthalpy value (20–40 kJ mol^{-1}) [30]. Due to the requirement for deficient activation energy, this process is considered reversible. The physicochemical properties of biogas components are reported in Table 4.1.

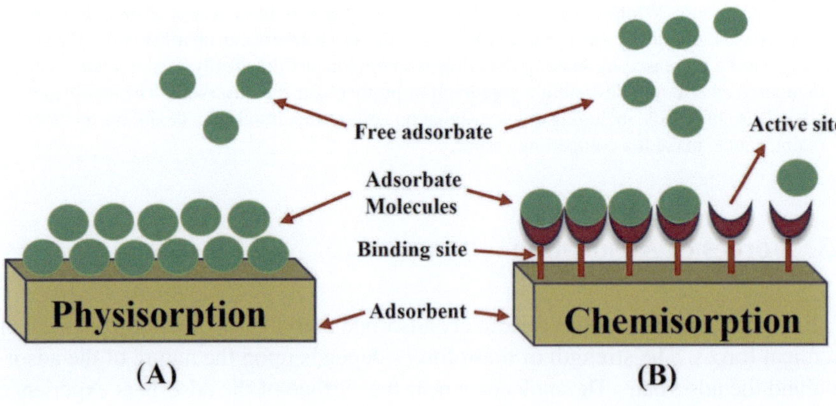

Fig. 4.1 **a** Physical adsorption, **b** chemical adsorption. Reproduced with Permission from Elsevier [19]

Table 4.1 Physiochemical properties of gaseous components [10, 40]

Gaseous component	σ [Å]	α [Å^3]	μ [D]	Θ [D-Å]	Tc [K]
CH_4	3.80	2.448	0.000	0.02	190
CO_2	3.30	2.507	0.000	4.30	304
H_2S	3.60	3.630	0.970	3.74	373
H_2O	2.65	1.501	1.850	2.30	647
N_2	3.64	1.710	0.000	1.54	126

σ kinetic diameter (Å: Angstrom), α polarizability (Å^3: Angstrom^3), μ dipole moment (D: coulombmeter), Θ: quadrupole moment, Tc: critical temperatures.

ii. **Chemical adsorption**: This process involves chemical forces of attraction or the formation of chemical bonds. These forces are strong enough to break and form new ionic or covalent bonds. It is a selective, irreversible, and exothermic process at high temperatures where the adsorbate molecules undergo electronic rearrangement with the adsorbent molecules [11]. With the increase in temperature, chemisorption first increases and then decreases. The enthalpy of this process is much higher (80–240 kJ mol^{-1}) than the physical adsorption process [30]. Chemisorption involves the formation of a unilayer of adsorbate on adsorbent with the formation of weak bonds. Since activation energy in chemical adsorption is high and takes place slowly, it is also called activated adsorption. Figure 4.1b represents chemical adsorption, where a weak bond is formed between the adsorbate and adsorbent molecule.

Pore size distribution

The separation performance of the adsorbent is determined by its porous texture. Pores are usually shaped into spherical pellets or extruded or shaped into monolithic honeycomb structures. The packing and piece density of the adsorbent are essential parameters in selecting the adsorbent. However, the pore diameter ratio to molecular diameter is essential rather than absolute pore size. Pores are categorized into micro-, meso-, and macropores according to their size, which is measured based on the diameter of cylindrical pores or the distance between the sides of slit-shaped pores in the IUPAC classification as shown in Fig. 4.2 [31].

4.4 Selectivity of Adsorbents

The basis of adsorption is the *selectivity* of the adsorbent for preferentially adsorbed species. This could be determined by the variation in the adsorption rates (kinetic selectivity) or the difference in the adsorption capacity of the gas components at equilibrium (equilibrium selectivity or thermodynamic selectivity). One or more of the following mechanisms may be used to accomplish the adsorption of components from a gas mixture [40].

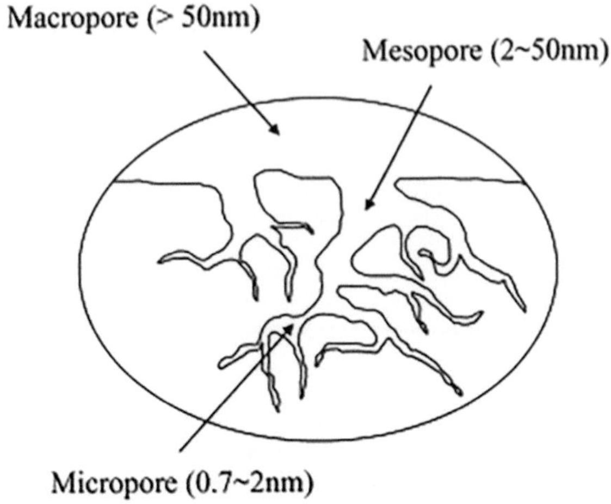

Fig. 4.2 Pore size distribution in adsorbents. Reproduced with Permission from Elsevier [13]

(a) ***Steric mechanism***: This mechanism is known as the molecular sieving effect in which one of the molecules of the gas mixture is not permitted to enter the adsorbent's pores due to its cross-sectional size or shape limitations. This mechanism is common in molecular sieves such as zeolites. In the case of separation of molecules having similar critical pore sizes, such as CO_2/CH_4 separation, molecular sieve-based processes are not likely to be viable [30].

(b) ***Thermodynamic equilibrium mechanism***: This is predicated on variations in the gas components' adsorption capacities at equilibrium. Strongly interacting molecules are held onto, while the others are released as effluent. Adsorbents based on equilibrium, such as silica gel, zeolite 13X, activated carbon, and metal–organic frameworks, exhibit robust surface contact and superior CO_2 selectivity. Because these adsorbents form stronger bonds with CO_2 than CH_4, they can absorb higher CO_2 loadings than CH_4 [40].

Equilibrium selectivity is one of the critical factors for calculating adsorbent separation capacity. Equilibrium selectivity can be measured by achieving system equilibrium in an unlimited period. For a given adsorbent and gaseous mixture, comprising X as the more strongly adsorbed component and Y as the less strongly adsorbed one, equilibrium selectivity ($\alpha_{X,Y}$) is generally represented by Eq. 4.1:

$$\alpha_{X,Y} = (n_X/n_Y)(p_X/p_Y) = K_X/K_Y \quad (4.1)$$

where p_X, p_Y are the component mole fractions in the gas mixture and n_X, n_Y is the equilibrium adsorption capacity of each gas component as estimated from the pure gas component isotherm. Henry's selectivity, also known as $\alpha_{X,Y}$, can be calculated as the ratio of X and Y Henry's constants (Ki). For PSA applications, this number between 2 and 104 is appropriate [40].

(c) ***Kinetic selectivity***: In the adsorption process, it is the main separation factor. The variations in the speeds at which molecules diffuse through the adsorbent pores are used to quantify it. It is selectivity that varies with time. While the slower-diffusing gaseous components concentrate in the effluent stream, the adsorbent retains the faster-diffusing ones. The adsorbent is chosen to prevent more massive CH_4 molecules with a kinetic diameter of 3.8 Å from diffusing through them. In comparison, smaller CO_2 molecules with a kinetic diameter of 3.4 Å can readily enter their structure [30]. Carbon molecular sieves, SAPO-30, titanosilicates, clinoptilolites, DDR zeolites, and carbon show comparable separation properties for CO_2 and CH_4. The ratio of mass transfer rates for the various components in the gas mixture is known as kinetic selectivity. The diffusivity coefficients $D_{X,Y}$ in the adsorbent pores are assumed to be proportional to ki. Therefore, the impacts of mass transfer coefficients for two components, kX, kY, are included in the kinetic selectivity factor $\eta_{X,Y}$. As a result, the equilibrium selectivity of the constituents in the gas mixture and the diffusivity ratio are connected to kinetic selectivity [30]

$$\eta_{X,Y} = \alpha_{X,Y}\sqrt{\frac{kX}{kY}} \quad \text{if } p_X = p_Y \text{ then; } \alpha_{X,Y}\left(\frac{nX}{nY}\right)\sqrt{\frac{kX}{kY}} \tag{4.2}$$

An adsorbent of high capacity with slow kinetics and an adsorbent of fast kinetics with low capacity are not preferred for efficient adsorption. Since adsorbent with slow kinetics requires long residence time, a low capacity requires more adsorbent material during adsorption in the column to achieve the desired product [30].

4.5 Classification of Adsorbents

The most crucial task in adsorption is the selection of a suitable adsorbent, followed by selecting fast-cyclic regeneration. Several adsorbents are available in different shapes, like spherical pellets, rods, moldings, or monoliths with a hydrodynamic radius of 0.25 and 5 mm for PSA. Adsorbents must have several characteristics to perform efficiently in an adsorption process.

The desirable characteristics for an adsorbent include:

(i) Adsorbent should have good adsorption capacity and high selectivity for the adsorbate even in the presence of moisture and other toxic contaminants.
(ii) It should have fast adsorption and regeneration kinetics.
(iii) Specific capacity.
(iv) Good chemical and physical strength throughout the adsorption and regeneration cycle.
(v) Fast inter-particle diffusion.
(vi) It should be able to regenerate at modest pressures or temperature swings.
(vii) Low cost per unit volume.
(viii) Low packing density and small pore diameters to avoid large columns.
(ix) High abrasion resistance, high thermal stability.

Among various adsorbents capable of separating CO_2 from gas mixtures, zeolite-based adsorbents and carbon-based adsorbents are the most commonly used adsorbents [36].

Recently, magnesium or copper-based metal–organic framework (MOF), silicalite, or silico-aluminophosphate sorbents (SAPOs) have gained attention due to high adsorption capacity due to high pore volume. Adsorbents are screened after analyzing equilibrium and kinetic data. Equilibrium adsorbents such as activated carbon, zeolite 13X, and metal–organic frameworks (MOFs) show stronger surface interactions with CO_2 than CH_4. Adsorbents based on kinetic separation, such as carbon molecular sieve 3K, have low loading rates of CH_4/CO_2 but faster gas diffusion rates of CO_2 than CH_4 in the micropores. According to Grande and Rodrigues [18, 17], equilibrium-based adsorbents perform less well in PSA than kinetic-based adsorbents. This is because equilibrium-based adsorbents' isotherms exhibit considerable nonlinearity. For biogas upgrading applications, both equilibrium and kinetic adsorbents are well-developed and commercially accessible. Table 4.2 lists some adsorbents' equilibrium and kinetic selectivity.

i. **Zeolites**

Zeolites are open structures of silica, also known as aluminosilicates. These are polar molecules intrinsically microporous with the general formula $[AlO_2)x(SiO_2)y]\cdot mH_2O$ [14]. Zeolites have a repeating tetrahedral crystalline pore network with uniform pore sizes and dimensions into which the molecules can penetrate.

The adsorption process in zeolites is based on the molecular sieving and polarity principle. Polar molecules such as H_2O, H_2S, SO_2, NH_3, etc., are strongly adsorbed

Table 4.2 Equilibrium and kinetic selectivity of adsorbents for CO_2 and CH_4 at 100 kPa [30]

Adsorbent	Absorbent name	Type	CO_2/CH_4 equilibrium selectivity	CO_2/CH_4 kinetic selectivity	References
Activated carbon (AC)	PCB, Calgon Corp	Equilibrium	3.29	–	[29]
Zeolite 5A	5A (Sinopec)	Equilibrium	5.19	3.6	[33]
Zeolite 13X	13X (Sinopec)	Equilibrium	8.92	4.5	[4]
Metal–organic frameworks (MOF)	Mg-MOF-74	Equilibrium	8.5	3.8	[4]
Activated carbon/carbon molecular sieves (AC/CMS)	CORK-DC-0	Kinetic	4.76	8.9	[6]
Metal–organic frameworks (MOF)	Cu-MOF	Kinetic	1.86	9.7	[3]

by zeolites. Zeolites exhibit good CO_2 adsorption capacity and selectivity at low pressures and moderate temperatures. However, due to their hydrophilic nature, moisture negatively affects their adsorption capacity and reduces the availability of the active surface area [25]. Most zeolites that are commercially available are produced synthetically. These have different properties and adsorption affinities for different compounds. Zeolites are commonly designated by letters and classified depending on pore size. The common zeolite-based adsorbents for gas separation applications are zeolite 4A, 5A, and 13X. Synthetic zeolite 13X is commonly used for the dehydration, desulfurization, and purification of natural gas and biogas. Natural zeolites, such as Clinoptilolite, Mordenite, Erionite, Ferrierite, and Phillipsite, are also available for gas separation. These are nonporous and low-cost zeolites. However, they have the disadvantage of the prior activation, processing, and shipping requirements, making them even more expensive. Therefore, natural zeolites must exhibit exceptional and better adsorption properties than synthetic ones. The characteristics of some zeolites are summarized in Table 4.3.

Figure 4.3 shows the adsorption equilibrium of CO_2 and CH_4 on zeolite 13X. The loading of CO_2 is greater than the loading of CH_4 at given P, T conditions. Additionally, the cyclic CO_2 capacity at lower temperatures is smaller than at higher temperatures, so zeolite 13X is more accessible to regenerate at 323 K than at 298 K [15].

Hernández-Huesca et al. [20] compared three different natural zeolites (erionite, mordenite, and clinoptilolite) as adsorbents for separating methane mixed with carbon dioxide and concluded that clinoptilolite exhibited better adsorption rates than others. Alonso-Vicario et al. [1] compared the performance of zeolite 5A and Clinoptilolite for upgrading and cleaning biogas. It was observed that clinoptilolite is the best material choice because of double adsorption capacity and complete regeneration and stability characteristics as compared to zeolite 5A.

Table 4.3 Characteristics of major zeolites [22, 35, 44]

Zeolite type	Property	
Synthetic	**Size A°**	**Bulk density, kg/m³**
3A	3	600
4A	4	615
5A	5	670
10X	8	660
13X	10	570
Natural	**Porosity [%]**	**Bulk density Kg/m³**
Clinoptilolite	34	1150
Mordenite	28	1170
Erionite	35	1510
Phillipsite	31	1580
Chabazite	47	1450

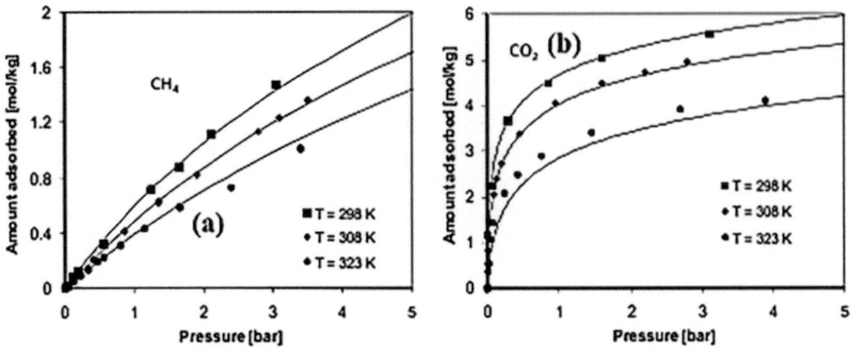

Fig. 4.3 Adsorption equilibrium of CH_4 and CO_2 on Zeolite 13X at different temperatures. Reproduced with Permission from Publisher

ii. **Carbon Adsorbents**

Carbon adsorbents are molecules with a porous structure containing small amounts of different heteroatoms, such as oxygen and hydrogen. This porous structure characterizes activated carbon into different types. Carbon adsorbents have low polarity, low heat of adsorption, less energy requirement for regeneration, and the capability of adsorbing non-polar and weakly polar organic molecules without the pre-requisite moisture removal.

Carbon adsorbents are classified into four categories:

- Activated carbons (ACs).
- Carbon molecular sieves (CMS).
- Activated carbon fibers (ACFs).
- Carbon-based nanomaterials.

Activated Carbons (ACs)

Activated carbon is a form of carbon processed to make microporous structures with high adsorption capacity, the strongest physical adsorption forces, and a high degree of surface reactivity and volume of adsorbing porosity. The presence of different surface functional groups, especially oxygen groups, causes adsorption [26] Carbonaceous raw resources such as wood, peat, coal, petroleum coke, bone, coconut shell, and fruit nuts are processed to create it. Activated carbon can be created primarily in two ways: chemically and physically. The physical activation method is a two-step process that begins with heating the raw material to a low temperature (about 773 K) without oxygen, carbonizing it into char. Next, oxidizing gas, such as air, carbon dioxide, or steam, activates the material at a high temperature (about 1273 K) [30]. The physical activation method produces AC with a fine pore structure. The chemical activation method includes impregnation and coating of a dehydrating chemical agent with alkaline or oxide solids, like sodium hydroxide, sodium carbonate, zinc chloride, sulfuric acid, or phosphoric acid, on the material before the heat treatment in an inert atmosphere. According to Mazlan et al. [24], this process yields activated carbon with a wide pore structure, which makes AC better suited for absorbing big molecules to raise the adsorption capacity and affinity for particular chemicals. Because regeneration is carried out at high temperatures, it can be intricate and energy-intensive. Because they are hydrophobic, activated carbons are often employed to adsorb low-polarity substances. Molecular sieves and activated carbons are the most commonly used carbon adsorbents for biogas upgrading.

As can be seen in Table 4.4, AC pores are classified into three primary groups: micropores, macropores, and mesopores. Among these, micropore ACs are the most effective adsorbent due to their large surface area and adsorption ability.

Carbon Molecular Sieves (CMSs)

Carbon molecular sieve (CMS) provides relatively high gas adsorption capacity and kinetic selectivity. Raw materials like polyvinylidene dichloride and phenolic resins or naturally occurring materials such as anthracite or hard coals are used for manufacturing CMS with effective diameters ranging from 0.4 to 0.9 nm [31]. The preparation of CMS is comparable to activated carbons but includes additional treatment with organic molecules that are polymerized with the carbonaceous raw material. CMS for large-scale applications is obtained from coal. CMS can be obtained from activated carbons by modifying the pore structure. The pour mouths are coated with a carbonized or coked thermosetting polymer with better kinetic properties and high selectivity. These microporous adsorbents have a bi-dispersed pore structure with clearly distinguishable macropores and micropores.

Carbon molecular sieves are commonly employed as an alternative to activated carbon and zeolites. Activated carbon and zeolites are equilibrium adsorbents based on the adsorption capacity compared to CMS, which are kinetic adsorbents. Carbon molecular sieves are best for biogas upgrading. CMS allows CO_2 to diffuse within the micropore network, whereas it presents resistance to methane. Cavenati et al. [7] studied the adsorption equilibrium isotherms of CO_2 and CH_4 in CMS-3K (Takeda Corp., Japan). CMS presented a kinetic selectivity toward CO_2.

Grande and Rodrigues [17] also compared the performance of CMS-3K and zeolite 13X in a five-step PSA cycle for biogas upgrading. Both adsorbents obtained 98% CH_4 purity, but CMS-3K achieved 33% higher CH_4 recovery with 50% less energy consumption than zeolite 13X. Grande and Rodrigues [17] compared the adsorption of CO_2 on zeolite 13X and carbon molecular sieve (CMS-3K). CMS upgraded 55% CH_4 in raw biogas to fuel-grade methane with 98% purity and

Table 4.4 Characteristics of activated carbons [23, 26, 31]

Characteristics of activated carbons	Micropores	Mesopores	Macropores
Diameter (nm)	< 2	2–50	> 50
Pore volume (cm^3/g)	0.15–0.5	0.02–0.1	0.2–0.5
Surface area (m^2/g)	100–1000	10–100	0.5–2
Particle density	0.6–0.9 g/cm^3		
Porosity	0.4–0.6		

presented a high trade-off between purity and recovery with less consumption of power than zeolite. However, in zeolite 13X, only a 60% recovery of CH_4 was achieved.

Using a two-column PSA, Cebula [9] investigated the behavior of carbon molecular sieve (CMS) KP-407 as a selective adsorbent for upgrading biogas. Significant and sufficient kinetic selectivity to CO_2 was demonstrated by CMS material, which could convert biogas to biomethane with pipeline quality (<97.5%) and recovery above 90% (93.8%). Tomadakis et al. [41] used 4A, 5A, and 13X molecular sieves in a series of pressure swing adsorption tests to investigate the separation of bulk volumes of H_2S from CO_2. All sieves showed high selectivity of H_2S over CO_2, with the 13X and 5A exhibiting the highest levels of selectivity. At high product recovery rates, pure CO_2 was generated during the adsorption step using new 5A and 13X sieves. Using both new 5A and regenerated 4A zeolites, effective H_2S purification was accomplished. The effectiveness of the 4A and 13X molecular sieves in the presence of moisture was compared by CO_2. The significant CO_2 adsorption capability of the 13X molecular sieve from dry biogas substantially enhanced the moisture content. When there is water present, bicarbonate ions mix with the water that has been adsorbed to generate the predominant type of CO_2 adsorption. On the other hand, when it is absent, various carbonate ions develop in conjunction with molecular adsorbed species. Because CH_4 is co-adsorbed, the adsorption capacity of CO_2 in the 4A molecular sieve is less than 13X. However, the presence of water has little effect on the adsorption capacity of CO_2 in molecular sieves.

iii. **Metal–Organic Framework**

Metal–organic framework (MOF) is formed by synchronizing bonds between metal salts and multidentate ligands. MOFs are promising materials with crystalline two- or three-dimensional porous structures. These are mainly synthesized with the advantages of zeolites. These materials have a high strength due to strong bonds between inorganic molecules and organic units. MOFs have a high pore volume and pore diameter which increases gas adsorption capacity. MOF successfully uses vacuum pressure swing adsorption (VPSA) units for CO_2/CH_4 separation for biogas upgrading. MOFs have the advantage of reduced power consumption and size of the adsorption units.

A variety of MOFs are efficient for CO_2 adsorption. Metal–organic framework extrudates such as Cu-BTC [polymeric copper (II) benzene-1,3,5-tricarboxylate] have higher adsorption characteristics than 13X zeolite. It has a high CO_2 adsorption capacity of 6.6 mol/kg at 2.5 bar and 303 K. Cavenati et al. [8] studied adsorption equilibrium data of CO_2/CH_4 on Cu-MOF extrudates which presented small nonlinearity of isotherm with high absorption capacity for CO_2 along with it. However, many problems, such as the effect of impurities like O_2, H_2O, H_2S, SOx, NOx, etc., present in the input gas stream on its adsorption behavior, process stability in multiple adsorption/desorption cycles, and mechanical stability, need to be further worked upon. Wu et al. [43] compared the adsorption performance of MOF 508b with traditional adsorbents such as zeolite 13X and carbon molecular sieve 3 K in a two-column six-step PSA system. The study revealed that the energy consumption by zeolite 13X is the highest compared to that of metal–organic framework 508b, 56% lower. MOF 508b presented a higher adsorption capacity than molecular sieve 3K. It has a linear CO_2 isotherm with easy desorption and a 13% smaller column diameter than zeolite 13X.

In conclusion, the presence of pollutants other than CH_4 and CO_2 in raw biogas, such as H_2S and H_2O, is important to consider when choosing adsorbent materials for biogas upgrading utilizing the PSA process. These contaminants can harm the adsorbent. Similar to carbon dioxide adsorption, the adsorption material also adsorbs hydrogen sulfide, but this process is irreversible. Therefore, eliminating H_2S before the PSA procedure becomes essential. In the same vessel as CO_2, H_2O can be removed by using carbonaceous materials, such as activated carbon. Zeolites, however, cannot remove H_2O concurrently because the water adsorption gradient is too steep, making desorption exceedingly challenging.

4.6 PSA-Based Biogas Upgrading Process

Pressure swing adsorption is a proven biogas upgradation method. It comprises 21% of biogas upgrading market. This process produces 96–98% pure biomethane. PSA may remove nitrogen and oxygen from the input gas stream using more complicated adsorbent bed setups, which increases plant upgrade costs. Before this process, biogas H_2S, moisture, and siloxane must be removed because they irreversibly damage the adsorbent material. Unfortunately, PSA's low CH4 recovery, especially compared to other biogas upgrading systems, is due to the off-gas loss [42]. An off-gas with a CH_4 percentage of roughly 15–20% is frequently produced, and a CH_4 recovery of 85–90% is typically achieved [27]. The high CH4 content of the off-gas requires further treatment to avoid its discharge [2]. According to Bauer et al. [5], the PSA-based upgrading plant has an availability that ranges from 94 to 96%, which may be quantified as the proportion of its entire operational time that it is really being utilized.

The energy needed to convert raw biogas to biomethane is significant when choosing a method. Less energy needed for upgrading means greater end-use energy. PSA plant compressors and vacuum pumps use electricity. Pressure swing adsorption requires electricity for biogas compression to 4–7 bar and dehydration. However, energy usage depends on plant size and case. This is 0.24–0.3 kWh/Nm^3 raw biogas and 0.29–0.6 kWh/Nm^3 improved biogas [28].

PSA is a dynamic process that runs in a loop. Adsorbent-filled vertical columns run sequentially under pressure with raw or improved biogas. Adsorption, pressure equalization, blowdown, and purge occur simultaneously in the columns. PSA columns at 4–10 bars are used to hold CO_2, N_2, O_2, and other chemicals in the adsorbent's pores [5]. By lowering the pressure, CH_4 can be collected from the top of the column where it flows unretained. The above processes are repeated in a cyclical manner. Within this sequence, commercial upgrading plants run four, six, or nine adsorber columns in tandem. Adsorbents that have been extensively loaded with CO_2 gas must be regenerated step-by-step. The Skarstrom cycle is this repeated pattern of adsorption and regeneration that typically lasts between two and ten minutes [15]. Bauer et al. [5] describe the four PSA cycles as:

The first step is adsorption, which involves supplying biogas into an adsorber at a bottom pressure of 6–8 bars after the water vapor and hydrogen sulfide have been removed. Gases such as CO_2, O_2, N_2, and others are absorbed preferentially by the medium as they flow through the pipe, with the gas escaping as methane. The process of transferring biogas to a different, previously regenerated vessel allows for continuing operation even after the adsorbent material is fully saturated.

The second step is depressurizing the adsorber tank to air pressure and then to near vacuum. The pressure of 6–8 bar is first decreased to 3–4 bar and subsequently to atmospheric pressure by pressure communication with column 4, which was degassed by a small vacuum.

Third, regenerate or purify the column by sucking its pressure to 0.1 bar (VSA). The majority of the desorbed gas is CO_2, with some CH_4. A purge stage extracts pure CH4 from the CH_4 product end to minimize CO_2 in the desorbed gas. Depressurization during regeneration alters off-gas composition. The off-gas contains a significant amount of CH_4 at high pressures, while most CO_2 is desorbed at low pressures. The off-gas with a high CH_4 concentration from the initial stage of decompression is delivered back to the raw biogas input to reduce CH_4 loss. The high CO_2 off-gas can be released into the environment, sent to the next adsorption step, or treated in an off-gas treatment unit after regeneration. The neighboring regenerated adsorption column receives CO2-saturated column gases. The column is pressurized while the previous one is progressively depressurized to atmospheric pressure. A mixture of carbon dioxide and methane with a high methane concentration is emitted and recovered into the system's input to reduce energy usage. Washing the saturated column with biogas regenerates the adsorbent material.

The final step is building up pressure, in which the purge is also done at low pressure; it is necessary to boost the pressure to start a new cycle. Before the adsorption phase resumes, the adsorber tank is slowly repressurized to the final adsorption pressure. After a pressure balance, an adsorber in its adsorption mode uses feed gas to achieve the final pressure build-up.

A typical design for PSA units comprises four columns since this cycle has four phases, as Fig. 4.4 illustrates. While the other three columns are in various stages of regeneration, one column is continuously involved in adsorption. In this setup, heavy (mainly absorbed) gas breaks through this column and connects to a second column (trim bed) where more heavy gas can be adsorbed [21].

Spoorthi et al. [38] examined the performance of PSAs with two beds, modified two beds, three beds, and four beds. With 99.9% product purity, the three-bed and modified two-bed PSA worked best. The three-bed PSA can reduce equipment size fourfold and energy usage twofold. Santos et al. [34] studied a six-step cycle operating at 323 K in a two-column PSA system. A decrease in CH_4 purity was observed when CO_2 was recycled in the purge and equalization steps. However, these steps increase the product recovery to more than 80%, obtaining methane with a purity of 97.1%, lost in the blowdown step.

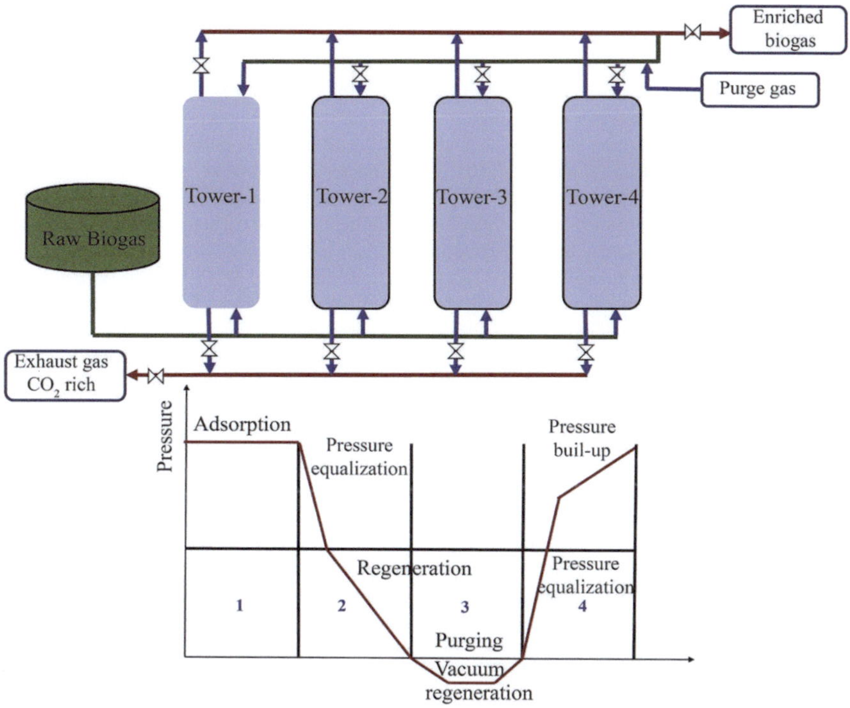

Fig. 4.4 Process diagram for upgrading biogas using PSA. Reproduced with Permission from Elsevier [37]

4.7 Conclusion

The primary goals of PSA research and development are to enhance low-energy technology, boost adsorbent efficiency, and reduce CH_4 losses. Sophisticated and potent adsorbent material development is one of the most crucial research directions for PSA technology. Therefore, one of the most important areas of research is the creation of novel adsorbents. These adsorbents include carbonaceous adsorbents, metal–organic frameworks (MOFs), zeolites, graphene, etc. The off-gassing from PSA and membrane separation processes contains a lot of methane. It is imperative that sorbents with high sorption and affinity, as well as separation competence, remove CO_2, H_2S, and other contaminants from biogas [40]. For the process to be lifetime, the material must also have strong chemical and thermal stability. In order to upgrade biogas, new alternative materials must be investigated for sorption (sorbents) and separation (membrane) technologies [43]. The material being used in the process determines minimal operating and maintenance costs as well as maximum separation efficiency for biogas upgrading via adsorption.

References

1. Alonso-Vicario A, Ochoa-Gómez JR, Gil-Río S, Gómez-Jiménez-Aberasturi O, Ramírez-López CA, Torrecilla-Soria J, Domínguez A (2010) Purification and upgrading of biogas by pressure swing adsorption on synthetic and natural zeolites. Microporous Mesoporous Mater 134(1–3):100–107. https://doi.org/10.1016/j.micromeso.2010.05.014
2. Augelletti R, Conti M, Annesini MC (2017) Pressure swing adsorption for biogas upgrading. A new process configuration for the separation of biomethane and carbon dioxide. J Clean Prod 140:1390–1398. https://doi.org/10.1016/j.jclepro.2016.10.013
3. Bao Z, Alnemrat S, Yu L, Vasiliev I, Ren Q, Lu X, Deng S (2011) Kinetic separation of carbon dioxide and methane on a copper metal-organic framework. J Colloid Interface Sci 357(2):504–509. https://doi.org/10.1016/j.jcis.2011.01.103
4. Bao Z, Yu L, Ren Q, Lu X, Deng S (2011) Adsorption of CO_2 and CH_4 on a magnesium-based metal organic framework. J Colloid Interface Sci 353(2):549–556. https://doi.org/10.1016/j.jcis.2010.09.065
5. Bauer F, Hulteberg C, Persson T, Tamm D (2013) Biogas upgrading—review of commercial technologies (Biogasuppgradering-Granskning av kommersiella tekniker) SGC Rapport 2013:270 "Catalyzing energygas development for sustainable solutions" www.sgc.se
6. Cansado IPP, Mourão PAM, Ribeiro Carrott ML, Carrott PJM (2010) Activated carbons prepared from natural and synthetic raw materials with potential applications in gas separations. Adv Mater Res 107(1):1–7. https://doi.org/10.4028/www.scientific.net/AMR.107.1
7. Cavenati S, Grande CA, Rodrigues AE (2005) Upgrade of methane from landfill gas by pressure swing adsorption. Energy Fuels 19(6):2545–2555. https://doi.org/10.1021/ef050072h
8. Cavenati S, Grande CA, Rodrigues AE, Kiener C, Müller U (2008) Metal organic framework adsorbent for biogas upgrading. Ind Eng Chem Res 47(16):6333–6335. https://doi.org/10.1021/ie8005269
9. Cebula J (2009) CEBULA-2009
10. Chaemchuen S, Kabir NA, Zhou K, Verpoort F (2013) Metal-organic frameworks for upgrading biogas via CO_2 adsorption to biogas green energy. Chem Soc Rev 42(24):9304–9332. https://doi.org/10.1039/c3cs60244c
11. Chiou CT (2002) Frontmatter. In: Partition and adsorption of organic contaminants in environmental systems. Wiley, London. https://doi.org/10.1002/0471264326.fmatter
12. Dabrowski A (2001) Adsorption from theory to practice. Advances in colloid and interface science, pp 135–224
13. Eom SW, Lee CW, Yun MS, Sun YK (2006) The roles and electrochemical characterizations of activated carbon in zinc air battery cathodes. Electrochim Acta 52(4):1592–1595. https://doi.org/10.1016/j.electacta.2006.02.067
14. Georgiev D, Bogdanov B, Angelova K, Markovska I, Hristov Y (2009). Synthetic zeolites—structure, classification, current trends in zeolite synthesis. In: Economics and Society Development on the Base of Knowledge: International Scientific Conference, vol 7
15. Grande CA (2012) Advances in pressure swing adsorption for gas separation. ISRN Chem Eng 2012:1–13. https://doi.org/10.5402/2012/982934
16. Grande CA, Ribeiro RPL, Oliveira ELG, Rodrigues AE (2009) Electric swing adsorption as emerging CO_2 capture technique. Energy Procedia 1(1):1219–1225. https://doi.org/10.1016/j.egypro.2009.01.160
17. Grande CA, Rodrigues AE (2007) Layered vacuum pressure-swing adsorption for biogas upgrading. Ind Eng Chem Res 46(23):7844–7848. https://doi.org/10.1021/ie070942d
18. Grande CA, Rodrigues AE (2007a) Biogas to fuel by vacuum pressure swing adsorption I. Behavior of Equilibrium and kinetic-based adsorbents. https://doi.org/10.1021/ie061341
19. Hasan HF, Al-Sudani FT, Albayati TM, Salih IK, Hharah HN, Majdi HS, Cata Saady NM, Zendehboudi S, Amari A, Gheni SA (2024) Solid adsorbent material: a review on trends of post-combustion CO_2 capture. Process Saf Environ Prot 182:975–988. https://doi.org/10.1016/J.PSEP.2023.12.025

20. Hernández-Huesca R, Díaz L, Aguilar-Armenta G (1999) Adsorption equilibria and kinetics of CO_2 , CH_4 and N_2 in natural zeolites. Separ Purif Technol 15
21. Kim YJ, Nam YS, Kang YT (2015) Study on a numerical model and PSA (pressure swing adsorption) process experiment for CH_4/CO_2 separation from biogas. Energy 91:732–741. https://doi.org/10.1016/j.energy.2015.08.086
22. Kusumastuti R, Sriyono, Pancoko M, Butar-Butar SL, Putra GE, Tjahjono H (2019) Study on the mechanism of CO_2 adsorption process on zeolite 5A as a molecular sieve in RDE system: an infrared investigation. J Phys Conf Ser 1198(3). https://doi.org/10.1088/1742-6596/1198/3/032009
23. Kwiatkowski JF (2011) Activated carbon: classifications, properties and applications. Retrieved from https://books.google.co.in/books?id=EuE9YgEACAAJ
24. Mazlan MAF, Uemura Y, Yusup S, Elhassan F, Uddin A, Hiwada A, Demiya M (2016) Activated carbon from rubber wood sawdust by carbon dioxide activation. Procedia Eng 148:530–537. https://doi.org/10.1016/j.proeng.2016.06.549
25. Moshoeshoe M, Silas Nadiye-Tabbiruka M, Obuseng V (2017) A review of the chemistry, structure, properties and applications of zeolites. Am J Mater Sci 2017(5):196–221. https://doi.org/10.5923/j.materials.20170705.12
26. Oubagaranadin J, Murthy ZVP (2011) Activated carbons: classifications, properties and applications, pp 239–266
27. Patterson T, Esteves S, Dinsdale R, Guwy A (2011) An evaluation of the policy and techno-economic factors affecting the potential for biogas upgrading for transport fuel use in the UK. Energy Policy 39(3):1806–1816. https://doi.org/10.1016/j.enpol.2011.01.017
28. Pertl A, Mostbauer P, Obersteiner G (2010) Climate balance of biogas upgrading systems. Waste Manage 30(1):92–99. https://doi.org/10.1016/j.wasman.2009.08.011
29. Ritter JA, Yang RT (1987) Equilibrium adsorption of multicomponent gas mixtures at elevated pressures. Ind Eng Chem Res 26. https://pubs.acs.org/sharingguidelines
30. Rufford TE, Smart S, Watson GCY, Graham BF, Boxall J, Diniz da Costa JC, May EF (2012) The removal of CO_2 and N_2 from natural gas: a review of conventional and emerging process technologies. J Petrol Sci Eng 94–95:123–154. https://doi.org/10.1016/j.petrol.2012.06.016
31. Ruthven DM (1984) Principles of adsorption & adsorption processes. AIChE J 31(3):523–524. https://doi.org/10.1002/aic.690310335
32. Ruthven DM, Farooq S, Knaebef KS (1994) Pressure swing adsorption
33. Saha D, Bao Z, Jia F, Deng S (2010) Adsorption of CO_2, CH_4, N_2O, and N 2 on MOF-5, MOF-177, and zeolite 5A. Environ Sci Technol 44(5):1820–1826. https://doi.org/10.1021/es9032309
34. Santos MPS, Grande CA, Rodrigues AE (2011) Pressure swing adsorption for biogas upgrading. Effect of recycling streams in pressure swing adsorption design. Indus Eng Chem Res 50(2):974–985. https://doi.org/10.1021/ie100757u
35. Santos MPS (2011) Advanced modelling of PSA processes for biogas upgrading. Master Thesis Aos meus pais…. Advanced Modelling of PSA Processes for Biogas Upgrading
36. Sarker MAI (2012) Equilibrium and mass transfer behaviour of CO_2 adsorption on zeolites, carbon molecular sieve, and activated carbons
37. Shah G, Ahmad E, Pant KK, Vijay VK (2021) Comprehending the contemporary state of art in biogas enrichment and CO_2 capture technologies via swing adsorption. Int J Hydrogen Energy 46(9):6588–6612. https://doi.org/10.1016/j.ijhydene.2020.11.116
38. Spoorthi G, Thakur RS, Kaistha N, Rao DP (2011) Process intensification in PSA processes for upgrading synthetic landfill and lean natural gases. Adsorption 17(1):121–133. https://doi.org/10.1007/s10450-010-9302-6
39. Sreeremya S (2017) Adsorption-review. Int J Adv Res Develop. www.ijarnd.com
40. Tagliabue M, Farrusseng D, Valencia S, Aguado S, Ravon U, Rizzo C, Corma A, Mirodatos C (2009) Natural gas treating by selective adsorption: Material science and chemical engineering interplay. Chem Eng J 155(3):553–566. https://doi.org/10.1016/j.cej.2009.09.010
41. Tomadakis MM, Heck HH, Jubran ME, Al-Harthi K (2011) Pressure-swing adsorption separation of H_2S from CO_2 with molecular sieves 4A, 5A, and 13X. Sep Sci Technol 46(3):428–433. https://doi.org/10.1080/01496395.2010.520292

42. Ullah Khan I, Hafiz Dzarfan Othman M, Hashim H, Matsuura T, Ismail AF, Rezaei-Dasht Arzhandi M, Wan Azelee I (2017) Biogas as a renewable energy fuel—a review of biogas upgrading, utilisation and storage. Energy Convers Manage 150:277–294. https://doi.org/10.1016/j.enconman.2017.08.035
43. Wu B, Zhang X, Xu Y, Bao D, Zhang S (2015) Assessment of the energy consumption of the biogas upgrading process with pressure swing adsorption using novel adsorbents. J Clean Prod 101:251–261. https://doi.org/10.1016/j.jclepro.2015.03.082
44. Yang RT (2003) Frontmatter. In: Adsorbents: fundamentals and applications. Wiley. https://doi.org/10.1002/047144409x.fmatter

Chapter 5
Water Scrubbing Technology

5.1 Introduction

Gas absorption is a mass transfer unit operation extensively applied in the chemical industry. During absorption, a gas mixture is brought into contact with a liquid to dissolve one or more constituents (solute) into the liquid (solvent). The rate of absorption of the solute from the gas mixture through the gas–liquid boundary (interface) is controlled by the diffusion rate. "Desorption" or "stripping" is the reverse of absorption. Mass moves in the reverse direction during desorption, also known as stripping, from the liquid phase to the gas phase. Both principles are similar and are often integrated into a system to separate the solute from an incoming gas mixture by absorption in a solvent and then recovering that solute in a more concentrated form from the loaded solvent for further use. The regenerated solvent is usually recycled to the absorber [11].

Gas absorption is carried out in vertical cylindrical columns called the packed bed scrubbing columns. Depending on the corrosiveness of the liquid streams and gas as well as the working process circumstances, the columns can be formed of steel, plastic, or a combination of these materials. The scrubbing column is filled with packing material. The liquid and gas flow currently over the packing material develops a large interfacial surface area for mass transfer and increases the intimate contact between the two phases [14]. Packed bed absorbers provide high removal efficiencies, handle high liquid rates, and have low solvent requirements compared to other gas absorbers. However, packed towers can also have high fouling and clogging potential, high drops in system pressure, and high maintenance costs. Installation, operation, and wastewater disposal add to the cost of the packed bed absorption systems.

V. K. Vijay et al., *Compressed Biogas (CBG)*, Green Energy and Technology,
https://doi.org/10.1007/978-981-96-5697-4_5

5.2 Design of Packed Bed Scrubbing Column

The packed bed scrubber is designed for maximum absorption of CO_2 in the solvent from a gas mixture. During the design of a packed bed scrubber, specific parameters are fixed by regulations or working conditions. Parameters like temperature, gas volume, and composition of the gas stream to be treated are usually known. The gas composition required at the outlet is usually fixed. The designer needs to identify (1) the best solvent; (2) the pressures at which the packed bed absorption column will operate; (3) temperatures of gas and solvent entering and departing the absorber; (4) the optimum velocity of gas through the absorber, i.e., the diameter of the vessel; (5) height of the vessel and its internal components, which is the type and height of the packing; (6) mechanical design of the absorption vessels, including flow distributors, packing supports, etc. The quality of the gas stream to be treated is generally known. This determines the temperature, volume, and composition of the gas stream. Emission standards must be met to help set the outlet composition of the gas stream. The absorbing liquid (solvent) and temperatures are also generally known [23].

The following are the main unknowns in the absorption system design:

- the liquid's flow rate,
- the vessel's required diameter to handle the liquid and gas flow, and
- the height of the absorber needed to achieve the desired removal.

5.3 Determination of Liquid Flow Rate

The law of mass conservation plays a pivotal role in determining the liquid flow rate for achieving high-efficiency gas absorption. The solute concentration in the gas and liquid phase is governed by a thermodynamic relation known as the equilibrium relation. At the equilibrium state, no solute transfer occurs from one phase to another. The extent of deviation from this state serves as a measure of the driving force for mass transfer.

According to the law of mass conservation, the material entering the process should either accumulate or exit. Material balance is a crucial tool in deciding the flow rates and compositions of the individual streams. In the context of gas absorption, the solute is the "material" in the material balance. The material balance in equation form is given by

$$-G_s * \mathrm{d}y = L_s * \mathrm{d}x; \quad (-\text{ve sign indicates the depletion of solute in the gas phase}) \tag{5.1}$$

where

L_s Flow rate of liquid on a free basis, kmol/h m^2
G_s Flow rate of gas on a solute-free basis, kmol/h m^2
x mole fraction of the solute in the liquid phase

y mole fraction of the solute in the gas phase
Y_1 molar ratio of gas at inlet
Y_2 molar ratio of gas at the outlet
X_1 molar ratio of liquid at the outlet
X_2 molar ratio of liquid at inlet

Assuming the total mass of the liquid and gas does not change, the operating line is obtained (defining the operating conditions within the absorber: what goes in and what comes out), and calculating on a solute-free basis, the equation becomes

$$L_s/G_s = (Y_1 - Y_2)/(X_1 - X_2) \tag{5.2}$$

Equation (5.2) is the operating line on the equilibrium diagram defining the operating conditions within the absorber. Its slope describes the liquid-to-gas ratio or the liquid's mass flow rate divided by the gas's mass flow rate. This equation passes through the points (X_1, Y_1) and (X_2, Y_2) and lies above the equilibrium line/curve since, during the absorption process, solute gets transferred from the gas phase to the liquid due to a higher concentration of solute in the gaseous phase than in liquid phase. As the slope increases, the specific consumption of liquid increases, which increases the driving force of mass transfer from a gaseous phase to the liquid phase [30].

In the packed bed absorption process, the flow rate of the liquid affects the gas removal efficiency. The packing's wetted surface increases with a high liquid flow rate, expanding the mass transfer's effective area. Because more liquid molecules are available to react with solute in a given amount of time, the gas–liquid boundary layer is continuously renewed, raising the liquid-side mass transfer coefficient. As the rate of liquid inflow increases, the liquid-film resistance decreases. Liquid flow rates are often at 25–100% above the minimum required quantity [15]. Generally speaking, the scrubber operation should be 50% higher than the minimum liquid flow rate (i.e., 1.5 times the minimum liquid-to-gas ratio).

5.4 Packing Material

Packing plays a vital role in gas absorption by facilitating an increased surface area for the liquid and gas phases to contact. Packing provides higher gas–liquid contact time during absorption, thus enhancing mass transfer efficiency [17]. Packing increases surface area for mass transfer and provides a sizeable wet surface for the gas. Higher column heights are, therefore, necessary to provide the necessary separation and purity.

The principal requisites of packing material are:

(i) strong but without excessive weight,
(ii) chemically inert to the fluids,
(iii) provides good contact between the liquid and gas,

(iv) have enough empty volume (passages) to accommodate both streams without experiencing an excessive liquid hold-up or pressure drop,
(v) judicious cost.

Packing material comes in various forms and sizes, each with unique characteristics, such as corrosion resistance, surface area, weight, pressure drop, and cost. Understanding these specific characteristics is critical to making informed decisions in gas absorption processes. The packing life depends upon the application, and under harsh conditions, packing life may be shorter due to corrosion, fouling, and breakage. This knowledge empowers us to make the best choices for our processes.

The most used packings for gas absorption are dumped or random (Raschig rings, Pall rings) and structured packings [13]. The knowledge of random packings—which are composed of designed shapes to maximize surface-to-volume ratio and minimize pressure drop—should reassure. These packings are available in various sizes, with smaller ones mainly used in pilot plant columns or laboratories. The units are more significant in stacked packings, ranging from 50 to 200 mm. The widespread use of random packings underscores their reliability and efficiency in gas absorption processes as shown in (Fig. 5.1).

The overall mass transfer coefficient in structured packing is more significant than that of random packing due to their larger surface area for flow distribution in gas–liquid contact. Structured packings are predominantly designed for complex separations. Random packings usually contain cheap, inert materials like plastics rather than steel or aluminum rings. The first random packings made especially for absorption towers were made of ceramic. Because ceramics are brittle, their use has decreased; nowadays, metal and plastic are more common. Acid gas and other highly corrosive pollutants are not packaged in metal. Applications involving high temperatures are not suitable for plastic packaging [4]. Table 5.1 lists the properties of commercial packings available. In the case of random packing, it is recommended that, if possible, the ratio of the diameter of the packing and column should be at least

a b

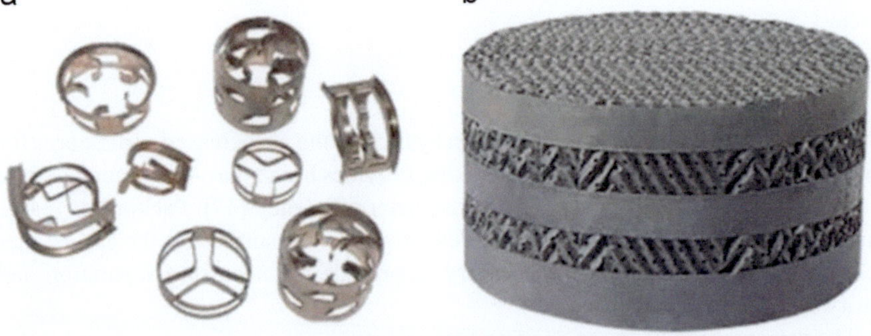

Fig. 5.1 Types of packing material (Left—random, right—structured). Reproduced with the permission from Elsevier [1]

$dp/D = 1/15$. Liquid distribution, and hence the mass transfer efficiency, depreciates rapidly above this size [28].

5.4.1 Determination of Packed Bed Height of Scrubbing Column (Z)

The depth of packing necessary to attain the intended removal efficiency is equal to the height of a packed column. It depends on how well the gas is separated and how pure the gas needs to be. As separation becomes more difficult, the packed bed gets taller. Because it impacts the scrubber's cost, efficiency, and rate of absorption, a correctly packed bed height is essential. The height of the packed bed is determined by multiplying (a) the number of required transfer units (NTU), a measure of the separation's difficulty (more transfer units are required for a higher level of product purity), by (b) the height of a transfer unit (HTU), a measure of the packings' effectiveness in separating different types of materials for different processes. The mass transfer coefficient impacts HTU. The value of HTU decreases with increasing mass transfer efficiency or with increasing mass transfer coefficient significance. It is generally acquired from experimental data. To ensure the highest accuracy, absorption equipment vendors usually conduct pilot plant investigations to find the HTU. When experimental data is unavailable, or only a preliminary estimation of absorber efficiency is required, generalized correlations are available in engineering texts such as *Chemical Engineers' Handbook* [10] or *Mass Transfer Operations* to calculate the height of a transfer unit.

5.5 Determination of Diameter of Scrubbing Column (D)

A scrubbing column's cross-sectional area and diameter are used to determine its capabilities. Suppose a packed bed column with a fixed gas and liquid flow rate. Reducing the column diameter results in an increase in the gas flow rate through the column. A state will be reached where the liquid flowing down the column becomes trapped in the vacuum spaces between the packing as the gas flow rate increases. The loading point is the ratio of gas to liquid at this moment. In this stage, the pressure drop in the column increases, and the degree of phase mixing decreases. As the velocity of gas rises, the liquid fills the vacuum spaces of the packing and finally forms a layer over the packing in the top section of the column [26]. The tremendous gas flows upwards in the column, causing no liquid to flow down. The pressure drop grows dramatically, ending the mixing and mass transfer between the phases. This condition is called flooding, and the gas velocity at which it happens is the flooding velocity. Hence, packed columns with a high diameter would not produce this problem. However, this adds to the expense of the column. Therefore, the packed

Table 5.1 Characteristics of commercial packings

Packing	Material	Size (mm)	Number (Pcs/m^{3})	Bulk density (kg/m^3)	Area m^2/m^3	Porosity (m^3/m^3) 100$^{\varepsilon}$	Packing factor (Fp)	References
Random packings								
Raschig ring	Ceramic	6	3,020,000	960	794	62	5250	[30]
		9	8,72,000	970	575	67	3280	
		12	3,77,000	880	368	64	2100	
		19	1,09,000	800	240	72	840	
		25	47,600	670	190	71	525	
	Metal	6	3,100,000	2130	787	72	2300	
		9	953,000	1500	508	81	1280	
		12	402,000	1200	417	85	980	
		19	117,000	830	276	89	605	
		25	50,000	620	207	92	375	
Pall rings	Metal	15	210,000	590	341	93	230	
		25	49,000	480	210	94	160	
		31	13,000	380	128	95	92	
		50	6000	350	102	96	66	
	Plastic	16	213,000	112	341	87	320	
		25	50,800	88	207	90	170	
		38	14,000	76	128	91	130	
		25	23,800	670	256	73	302	
Structured packings								
Flexipac	Steel	1	–	–	558	91	108	[10]
		2	–	–	223	93	72	
		3	–	–	134	96	52	
Flexiramic	Ceramic	28	–	–	282	70	131	
		48	–	–	157	74	79	
		88	–	–	102	85	49	
Intelox	Steel	1 T	–	–	315	95	230	
		2 T	–	–	213	97	160	
		3 T	–	–	177	97	92	
Suzlar	Metal gauge	AX	–	–	250	95	–	
		BX	–	–	492	90	60	
		CY	–	–	700	85	–	
		DX	–	–	900	78	–	

column's diameter is selected to operate at a fraction of the flooding velocity. The average operating range for the gas through the column is 50 to 75 percent of the flooding velocity [30].

In order to ensure proper liquid and gas distribution, packed bed columns are often engineered to operate at the lowest possible pressure drop. The random packing pressure drop should typically not exceed 80 mm of water per meter of packing height. The gas velocity will be approximately 80% of the flooding velocity at this number. For the absorbers, design values ranging from 15 to 80 mm of water per meter of packing are recommended.

In columns filled with random packings, the packing density is high near the wall of the column. Since the wall is a continuous surface and the voidage near it is high, it channels the liquid toward the walls and creates a path for the gas flow in the center of the column. Channeling is less prominent in columns, with the diameter of the individual packing piece being smaller than at least 1/8 of the column diameter. A ratio of packing diameter and column diameter as $dp/D = 1/15$ is usually recommended. Above this ratio, the liquid distribution and the mass transfer efficiency depreciate. Channeling is more severe in columns filled with structured packing, so it is not commonly used.

In actuality, the value for a given packing will rely on the column's diameter and height as well as the uniformity of the liquid distribution throughout, in addition to the physical characteristics and flow rates of the liquid and gas. The recommended packing size is less than 25 mm, and gas flow rate is less than 50 for less than 0.3 m packed bed column diameter. For large columns, 50 mm is often the greatest size of random packing that is advised.

5.6 Scrubbing Column Internals

For the best packing performance and effective mass transfer, it is crucial to choose the right scrubbing column internals, such as packed bed supports, liquid distributors and collectors, and other column internals. On the other hand, there is a lack of experimental evidence regarding the best internal cleaning of columns to improve mass transfer.

5.7 Water Scrubbing-Based Biogas Upgrading Process

5.7.1 Introduction

Water scrubbing is a simple, environment-friendly, cost-effective, and extensively used method for biogas cleaning and upgrading [29]. Among the commercially available technologies, it accounts for approximately 41% of the global biogas upgrading market [32]. The water scrubbing-based biogas upgrading process uses water as the solvent for physically absorbing gases like CO_2 and H_2S. Water as a solvent for gas absorption in packed bed columns is non-toxic, non-flammable, non-volatile, non-corrosive, non-viscous, non-foaming, chemically stable, and has a low freezing point. A significant benefit of using water as a solvent is its availability at low cost. Another benefit of using water is its regeneration, which does not require energy or specialized equipment [2]. These factors make water a better choice over chemical and organic solvents, which are costly and require energy for regeneration. With significant solubility for polar gases and low solubility for non-polar gases, it is the most prevalent solvent and is commonly referred to as a universal solvent.

The water scrubbing process is highly efficient based on the physical absorption of gases in water. CO_2 and H_2S are highly soluble in water at high pressure, while CH_4 is sparingly soluble. The difference in solubility for CH_4 and CO_2 in water is the driving force of the process. Theoretically, at STP, the solubility of CO_2 in water (0.609 × 10–3 mol fraction) is around 26 times higher than CH_4 (0.277 × 10–4 mol fraction) . This difference increases as process pressure rises, allowing CO2 to be absorbed more quickly and in higher amounts. At 25 °C, CH_4's water solubility is 73 times lower than that of H_2S and CO_2, respectively [8, 10]. Because H_2S is more soluble in water, it is thus extracted from biogas along with CO_2. This excellent efficiency should give people faith in the water scrubbing method's efficacy.

With decreasing temperatures, the solubility of gases increases; hence, temperature plays a significant role in the water scrubbing process. Lower temperatures in the absorption column are beneficial for increasing CO_2 absorption in water, not only because the solubility of CO_2 is high at low temperatures but also because it absorbs the process of dissolving heat released during absorption.

Läntelä et al. [16] saw a drop in CO_2 removal efficiency at 25 bar from 88.9 to 87.3% when the temperature rose from 10–15 to 20–25 °C. As the temperature rose from 7 to 40 °C, [33] observed a drop in CO_2 removal effectiveness from 85.3 to 52.2%. Consequently, it is important to pay closer attention to the temperature in the water washing column, particularly while the plant is operating in the summer.

H_2S gets dissolved in water simultaneously with CO_2 during the absorption process. The water scrubbing process can easily tolerate H_2S concentration of 300–2500 ppmv in biogas, a type of gas produced from the breakdown of organic matter in the absence of oxygen, but higher concentrations are harmful to the scrubbing equipment and the process [18]. H_2S dissolution in water decreases its pH. This leads to lesser solubility of CO_2 in water and clogging of the packing material because of microbial growth, which limits gas–liquid mass transfer and absorption. During the regeneration of water in the desorption/flash tank, H_2S, and CO_2 are released from water into the atmosphere, which causes an unpleasant odor and is also harmful to the environment. For this reason, it is highly recommended that H_2S be removed before the scrubbing process. Continuous renewal of water and usage of anti-foaming agents or the addition of alkaline material is required to avoid H_2S corrosion, poisoning, and fouling.

When designing a packed column for biogas upgrading, certain input parameters such as gas flow rate, working pressure, temperature, and the type of packing are fixed. A packed column, in this context, refers to a vertical vessel filled with a packing material that provides a large surface area for gas–liquid contact. The mathematical calculation for determining the height and diameter of the scrubbing column can yield deviations in results due to the unique nature of each case and the dependence on numerous variables. There is no completely satisfactory method for calculating the column's diameter and height. Therefore, the best method is to perform detailed experiments for a particular capacity and particular purification efficiency of packed columns and extrapolate the results for higher capacities with some error. However, estimations should be based on experimental values acquired from operating columns.

In practice, the column's diameter and the type and size of random packing are set for a particular gas flow rate. Vijay (1989) and Chandra et al. [7] used ceramic reaching rings for CO_2 removal in water scrubbing-based packed bed columns. Pall rings were used by Gamba and Pellegrini [9], Läntelä et al. [16], Rasi et al. [21, 22] in the packed bed column for CO_2 removal from biogas. Studied a biogas upgradation system based on water scrubbing technology using three packs of randomly packed plastic hollow spheres, zeolite pellets, and clay rings.

Water scrubbing facilities typically have a height-to-diameter ratio of 10:1 to 20:1 and water flow rates of 0.04–2 m^3/min per m^2 of the column area [30]. Rasi et al. [21] produced upgraded biogas with over 90% CH_4 content using a small height-to-diameter ratio of 3:1 (height 1.85 m, diameter 0.6 m), low water flow rates of 5 and 10 l/min (0.3–0.6 m^3/h), and high pressures of 10 to 30 bar. These methods replaced the conventional diameter/height ratio. The decrease of column height was made up for by the increased pressures and reduced water flow rate. [16] employed a 3:1 height-to-width ratio in addition to water flow rates of 5 l/min (0.3 m^3/h) and 11 l/min (0.66 m^3/h). They compensated for the small height in the absorption column by applying high pressures of 20–25 bar while recycling water, and they found that the highest carbon dioxide removal efficiency was achieved at 25 bar, at 88.9%. When compared to traditional plants, the low height-to-width ratio was blamed for the low CO_2 removal effectiveness.

Chandra et al. [7] employed a packed bed column with a diameter of 0.15 m and a height of 3.5 mm. Using a height-to-diameter ratio of 20:1, 93.4% of CO_2 was removed at 10 bars of pressure. Used a height-to-diameter ratio of 10:1 (1.0 m height and 0.1 m diameter) and achieved 73.14% CO_2 removal efficiency at 3 bar pressure. Therefore, it can be concluded that the packed bed column height and the packing type regulate the separation effectiveness in the packed column. In contrast, the diameter controls the gas quantity throughput. Consequently, a wider column can treat a bigger volume of biogas, while a higher bed may extract more CO_2 from biogas with lower incoming biogas flow rates.

5.7.2 Biogas Upgrading Process

The water scrubbing biogas upgrading process is a well-recognized and developed technology. Being a simple, environmentally friendly, and cost-effective method, it has a 41% share in the biogas upgrading market [32]. This technology can produce biomethane with a CH_4 purity of 95–99%. Upgrading plant developers usually ensures a maximum of 2% methane loss from water scrubbing plants. CH_4 loss of 1–2% has been reported by and Pertl et al. [19]. High CH_4 losses of 8% to 10% are also observed in some plants due to non-optimized system operation, which could include factors such as inefficient maintenance, inadequate monitoring, or suboptimal design [12]. CH_4 loss is dependent upon several factors. High flash pressure and water flow rates contribute to CH_4 loss from the system. However, the value of CH_4 loss sometimes rises due to temporary technical problems or different technical layouts.

A significant advantage of water scrubbing technology is its ability to remove H_2S simultaneously and CO_2 from biogas. This method can tolerate H_2S concentrations between 300 and 2500 ppm. H_2S dissolution in water decreases the pH. This, in turn, reduces the water solubility of CO_2 and causes corrosion and clogging of the packing material due to microbial growth. This results in partial gas–liquid mass transfer. Absorbed H_2S is released during the desorption process. The released gas stream has a high concentration of H_2S and must be treated before it is vented into the atmosphere to avoid health and environmental problems. Thus, prior removal of H_2S is recommended. Continuous regeneration of water and use of anti-foaming agents or the addition of alkaline salts is required to avoid H_2S corrosion, poisoning, and fouling [5, 31]. If the H_2S concentrations are above 2500 ppm, then it is a prerequisite to remove H_2S and moisture in order to avoid irreversible harm to the equipment. An acute problem in the water scrubbing column is microbial growth due to high temperature and the use of sewage water, which contains more nutrients and COD than normal process water. To overcome this problem, biocides (chemical substances that kill or inhibit the growth of microorganisms) are added, and the scrubber column is frequently cleaned.

The energy required to upgrade raw biogas to biomethane is vital for selecting technology. The lower the energy necessity for upgrading, the more net energy is obtainable for end use. Energy consumption during the water scrubbing-based biogas upgrading process is for the compression of raw gas and upgrade biogas (biomethane), water pumping, and cooling water and gas [6]. The amount of energy expended by these components depends on the composition of the raw biogas and the operation of the water scrubber. Generally, the energy requirement of a water scrubbing biogas upgrading system is 0.25–0.4 kWh/Nm^3 raw biogas. For example, Bauer and Hulteberg [5] stated that the specific electricity consumption is 0.23 kWh/Nm^3 when the plant capacity is 2000 Nm^3/h, but it rises to 0.3 kWh/Nm^3 if the capacity decreases to 400 Nm^3/h.

The energy required for compression of raw biogas at around 6–8 bar(a) pressures is usually between 0.10 and 0.15 kWh/Nm^3. The energy demand for pumping water depends upon the inlet and outlet pressure, the water volume, and the pump's efficiency [27]. The pump efficiency is nearly 80% at the whole load and 10–30% lower at the half load. However, the specific energy consumption increases considerably when operating at a lower load. Typically, 0.05–0.10 kWh/Nm^3 of energy is needed to pump the water at full load. Usually, 0.01–0.05 kWh/Nm^3 is required for cooling water and compressed gas. It depends upon the local conditions and the design of the water scrubber [20].

The capital, operational, and maintenance costs are crucial in determining the best biogas upgrading technology. However, inexpensive technology is not always a good choice. The capital and operational expenses of biogas upgrading technologies mostly depend on the product quality, composition of raw biogas, instrumentation, valves, and fittings, and notably on the plant capacity. An upgrading plant's operating and maintenance costs comprise electricity consumption costs, labor and chemicals or water costs depending on the specific upgrading technology used, and maintenance methods [24]. The plant's capacity considerably determines the investment cost of water scrubbing plants and decreases with the increase in throughput capacity. For a low-capacity plant, more or less the same amount of equipment, control valves, sensors, and pipes are required than for a much larger-capacity plant. If there is supplementary equipment and specific requirements, the investment cost increases. For plants in the operational range of 100 to 500 Nm^3/h, investment costs decrease from 10,100 to 3500 €/Nm^3/h, and it remains relatively constant for plant capacities over 1000 Nm^3/h in 1800–2000 €/Nm^3/h [18].

Initially, raw biogas is compressed at 8–12 bars. It is injected into the scrubbing column from the bottom. The scrubbing column is packed with randomly filled packing material. Water is directed from the column's top. The countercurrent interaction between them over the packing material increases the time that gas and water are in contact, which results in high mass transfer and increased H_2S and CO_2 absorption into water [5]. As the gas travels upwards in the column, more and more CO_2 gets absorbed in water, and a biomethane with above 90% CH_4 purity is achieved from the top. Water with absorbed gases is channeled to a flash vessel, where gases are desorbed at decreased flash pressures. CO_2-rich gas mixture (80–90% CO_2, 10–20% CH_4, and H_2S) is released from the flash vessel, which can be further processed for CH_4 loss recovery and bio-CO_2 production. It is noted that N_2 and O_2 cannot be separated since they are non-condensable gases [25].

Water is required in large quantities to upgrade biogas, but it depends upon the operating column pressure and gas temperature. Depending upon the quality, quantity, and cost of available water, biogas can be upgraded in two ways: single-pass scrubbing and regenerative process [3]. In single-pass scrubbing, wastewater from treatment plants is used. This method uses water only once, and the spent water obtained after the process is depressurized in a flash vessel and dumped into the drain, as shown in Fig. 5.2.

In the regenerative process, as shown in Fig. 5.3, spent water is regenerated in two steps. After depressurizing spent water in a flash vessel to desorb gases at two to three bar pressures, water is regenerated at atmospheric pressure in a desorption/stripping column. Scrubbing is done with recycled and repurposed regenerated water.

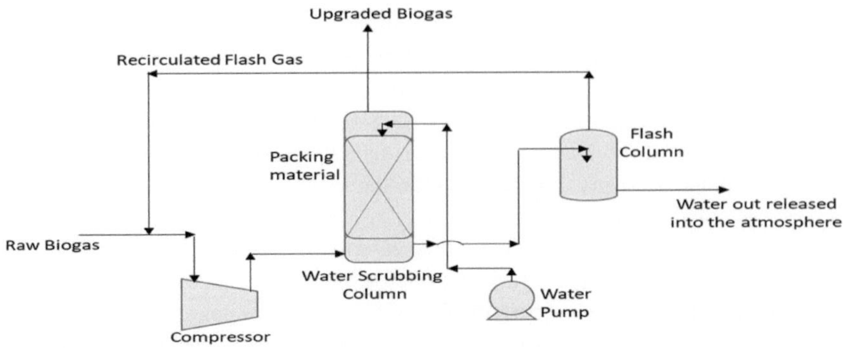

Fig. 5.2 Process flow diagram of biogas upgrading by water scrubbing process without water regeneration. Reproduce with Permission from Publisher [11]

In contrast to regenerative absorption units requiring a water flow rate of 0.18–0.23 m^3/Nm^3 biogas, single-pass biogas upgrading processes require 0.1–0.2 m^3/Nm^3 biogas [18]. In the regenerative water scrubbing process, water is recycled, but still, for biogas flow rates of 100–1000 Nm^3/h, 20–200 L/h fresh water is purged to avoid the accumulation of harmful compounds. The requirement for water depends upon operating pressure and temperature. High pressure requires less water for absorption. However, this increases the cost and energy requirement for biogas and water pumping compression.

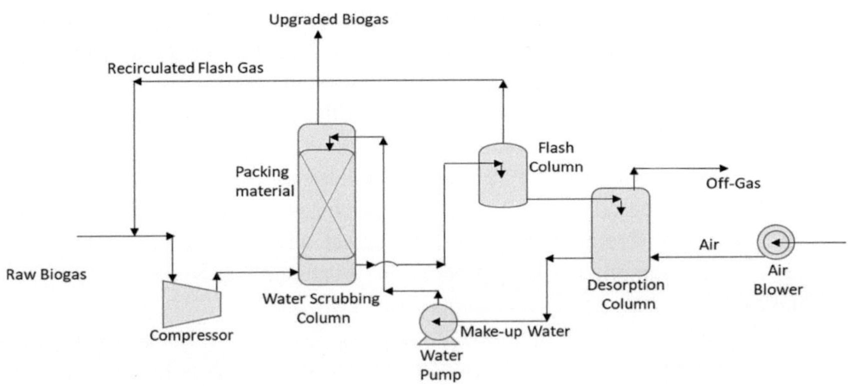

Fig. 5.3 Process flow diagram for upgrading biogas using water scrubbing and regeneration. Reproduce with Permission from Publisher [11]

5.8 Conclusion

Water scrubbing is a developed, easy, and simple technology increasingly acknowledged and employed worldwide. A comprehensive understanding of the variables influencing the effective extraction of CO_2 from biogas and the production of biomethane facilitates the design of packed bed columns for biogas upgrading. This method is capable of upgrading biogas to about 95% methane content. With further efficiency and cost reduction improvements, a more widespread technology implementation can be experienced, especially in developing countries.

References

1. Abdel-Salam AH, Simonson CJ (2016) State-of-the-art in liquid desiccant air conditioning equipment and systems. Renew Sustain Energy Rev 58:1152–1183. https://doi.org/10.1016/J.RSER.2015.12.042
2. Allegue LB, Hinge J, Allé K (2012) Biogas and bio-syngas upgrading
3. Angelidaki I, Treu L, Tsapekos P, Luo G, Campanaro S, Wenzel H, Kougias PG (2018) Biogas upgrading and utilization: current status and perspectives. Biotechnol Adv 36(2):452–466. https://doi.org/10.1016/j.biotechadv.2018.01.011
4. Arachchige USPR, Melaaen MC (2012) Selection of packing material for gas absorption. Eur J Sci Res 87. http://www.europeanjournalofscientificresearch.com
5. Bauer F, Hulteberg CP (2014) Biogas upgrading—review of commercial technologies. https://www.researchgate.net/publication/235932569
6. Budzianowski WM, Wylock CE, Marciniak PA (2017) Power requirements of biogas upgrading by water scrubbing and biomethane compression: comparative analysis of various plant configurations. Energy Convers Manage 141:2–19. https://doi.org/10.1016/j.enconman.2016.03.018
7. Chandra R, Vijay VK, Subbarao PMV (2012) Vehicular quality biomethane production from biogas by using an automated water scrubbing system. ISRN Renew Energy 2012:1–6. https://doi.org/10.5402/2012/904167
8. Cozma P, Wukovits W, Mămăligă I, Friedl A, Gavrilescu M (2013) Analysis and modelling of the solubility of biogas components in water for physical absorption processes. Environ Eng Manage J 12(1):147–162. https://doi.org/10.30638/eemj.2013.017
9. Gamba S, Pellegrini LA (2013) Biogas upgrading: analysis and comparison between water and chemical scrubbings. 32. www.aidic.it/cet
10. Green DW (2008) Perry's chemical engineers' handbook
11. Kapoor R, Ghosh P, Kumar M, Vijay VK (2019) Evaluation of biogas upgrading technologies and future perspectives: a review. Environ Sci Pollut Res 26(12):11631–11661. https://doi.org/10.1007/s11356-019-04767-1
12. Khan IU, Othman MHD, Hashim H, Matsuura T, Ismail AF, Rezaei-Dasht Arzhandi M, Azlee IW (2017) Biogas as a renewable energy fuel—a review of biogas upgrading, utilisation and storage. Energy Convers Manage 150:277–294. https://doi.org/10.1016/j.enconman.2017.08.035
13. Koch-Glitsch. (2010). Mass Transfer Products. http://www.kochglitsch.com/masstransfer/pages/Products.aspx. Assessed 14 Feb
14. Kolev N (2006) Packed bed columns: for absorption, desorption, rectification and direct heat transfer. Elsevier
15. Krehenwinkel H, Knapp H (1987) Pressure drop and flooding in packed columns operating at high pressures. Chem Eng Technol 10(1):231–242. https://doi.org/10.1002/ceat.270100128

16. Läntelä J, Rasi S, Lehtinen J, Rintala J (2012) Landfill gas upgrading with pilot-scale water scrubber: performance assessment with absorption water recycling. Appl Energy 92:307–314. https://doi.org/10.1016/j.apenergy.2011.10.011
17. Mendes MF (2011) HETP evaluation of structured and randomic packing distillation column. www.intechopen.com
18. Muñoz R, Meier L, Diaz I, Jeison D (2015) A review on the state-of-the-art of physical/chemical and biological technologies for biogas upgrading. Rev Environ Sci Biotechnol 14(4):727–759. Springer, Netherlands. https://doi.org/10.1007/s11157-015-9379-1
19. Pertl A, Mostbauer P, Obersteiner G (2010) Climate balance of biogas upgrading systems. Waste Manage 30(1):92–99. https://doi.org/10.1016/j.wasman.2009.08.011
20. Pöschl M, Ward S, Owende P (2010) Evaluation of energy efficiency of various biogas production and utilization pathways. Appl Energy 87(11):3305–3321. https://doi.org/10.1016/j.apenergy.2010.05.011
21. Rasi S, Läntelä J, Rintala J (2014) Upgrading landfill gas using a high pressure water absorption process. Fuel 115:539–543. https://doi.org/10.1016/j.fuel.2013.07.082
22. Rasi S, Läntelä J, Veijanen A, Rintala J (2008) Landfill gas upgrading with countercurrent water wash. Waste Manage 28(9):1528–1534. https://doi.org/10.1016/j.wasman.2007.03.032
23. Richardson JF, Harker JH, Backhurst JR (2002) Chemical engineering, vol 2 fifth edition, particle technology and separation processes
24. Rotunno P, Lanzini A, Leone P (2017) Energy and economic analysis of a water scrubbing based biogas upgrading process for biomethane injection into the gas grid or use as transportation fuel. Renewable Energy 102:417–432. https://doi.org/10.1016/j.renene.2016.10.062
25. Sahota S, Shah G, Ghosh P, Kapoor R, Sengupta S, Singh P, Vijay V, Sahay A, Vijay VK, & Thakur, I. S. (2018). Review of trends in biogas upgradation technologies and future perspectives. Bioresour Technol Rep 1:79–88. Elsevier Ltd. https://doi.org/10.1016/j.biteb.2018.01.002
26. Sherwood TK, Shipley GH, Holloway LFA (1938) Flooding velocities in packed columns. https://pubs.acs.org/sharingguidelines
27. Starr K, Gabarrell X, Villalba G, Talens L, Lombardi L (2012) Life cycle assessment of biogas upgrading technologies. Waste Manage 32(5):991–999. https://doi.org/10.1016/j.wasman.2011.12.016
28. Strigle RF (1987) Packed tower design and applications—random and structured packings, 2nd ed
29. Sun Q, Li H, Yan J, Liu L, Yu Z, Yu X (2015) Selection of appropriate biogas upgrading technology-a review of biogas cleaning, upgrading and utilisation. Renew Sustain Energy Rev 51:521–532. https://doi.org/10.1016/J.RSER.2015.06.029
30. Treybal RE (1980) Mass transfer operations, 3rd ed
31. Tynell Å, Börjesson G, Persson M (2007) Microbial growth on pall rings. Appl Biochem Biotechnol 299
32. UNIDO, United Nations Industrial development organization (2017) Biogas to biomethane, Fachverband Biogas, German Biogas Association. https://www.biogas-to-biomethane.com
33. Xiao Y, Yuan H, Pang Y, Chen S, Zhu B, Zou D, Ma J, Yu L, Li X (2014) CO_2 removal from biogas bywater washing system. Chin J Chem Eng 22(8):950–953. https://doi.org/10.1016/j.cjche.2014.06.001

Chapter 6
Physical Scrubbing

6.1 Introduction to Organic Scrubbing

Physical scrubbing is a well-established technology used for CO_2 separation from other gases. It is a simple process based on the solubility of CO_2 in the physical solvent according to Henry's Law, i.e., it depends upon pressure and temperature. The solubility of CO_2 increases as temperature drops and pressure rises. Therefore, low temperature and high pressure are ideal for CO_2 absorption. This process involves no chemical reaction between CO_2 and the solvent and is comparable to water scrubbing. Usually, the physical absorption process is considered when the CO_2 partial pressure is more than 3.5 bar and bulk removal of CO is required [2]. Depending on their solubility, other components in the gas mixture also get absorbed in the solvent during the process. Regeneration of spent solvent can be done by reducing pressure, inert gas stripping, or thermal regeneration. Minimum energy is required for pressure reduction and inert gas stripping; even if thermal regeneration could use a little bit more energy, it is still far less than other techniques. The physical absorption process's main benefit is relatively cheap startup and operational expenses. In addition, heat energy is not required for the absorption or regeneration process [13]. Additionally, this method is able to absorb H_2S selectively due to its higher solubility than CO_2. The only disadvantage of the physical absorption process is energy consumption for gas compression and solvent pumping for recycling [10].

6.2 Solvents for Physical Scrubbing

Several physical solvents are available for the capture of $CO_{2.}$ These require simply carbon steel construction for the process equipment and are relatively non-toxic and non-corrosive [4]. The use of organic solvents results in reduced absorbent recycling rates and compact units, which reduces investment and operating costs.

V. K. Vijay et al., *Compressed Biogas (CBG)*, Green Energy and Technology,
https://doi.org/10.1007/978-981-96-5697-4_6

i. Dimethyl Ether of Polyethylene Glycol (DEPG) (Seloxol Process)

DEPG, commonly known as selexol, is a blend of polyethylene glycol (CH_3O $(C_2H_4O)nCH_3$; n can range from 2 to 9). Since the 1960s, the gas separation industry has extensively employed and successfully validated the SelexolTM technology. It is popularly used for acid gases (CO_2 and H_2S) absorption in numerous applications due to its high capacity, solubility, and selectivity for acid gases. DEPG is appropriate for operation at temperatures between 0 °F (– 18 °C) and 347 °F (175 °C) because it has high viscosity at low temperatures [4]. This reduces the rate of mass transfer and increases packing requirements. Since it possesses low vapor pressure and inert chemical and thermal characteristics, the need for solvent recovery, purging, or makeup due to solvent loss is eliminated during the process. The low capital and operating costs and the heat requirement for spent solvent regeneration add to the advantages of Selexol [14]. Numerous businesses, notably Coastal Chemical Company (doing business as Coastal AGR), Dow (Selexol), and UOP (Selexol), hold licenses to produce and manufacture DEPG solvents. Under the Genosorb brand, clariant solvents are a class of di-alkyl ethers of polyethylene glycol [11].

ii. Methanol (MeOH) (Rectisol)

The primary advantage of methanol is its high selectivity for H_2S over CO_2. It has higher solubility of H_2S and CO_2 in methanol than DEPG [18]. It can separate H_2S and CO_2 from the gas mixture to levels as low as 0–0.1 ppm. Commercial processes based on chilled methanol as an absorbent are called the Rectisol process. In this process, raw biogas is cooled before being introduced into the absorption column. Low-temperature conditions in the range of – 40 °F to – 100 °F result in high solubility and reduced solvent losses due to low vapor pressure [4]. Methanol minimizes losses because it is totally miscible with water and does not froth. Its properties include strong chemical and thermal stability, non-corrosiveness, and non-degradability. The spent solvent can be easily regenerated by flashing at low pressure, eliminating the need for heat using a reboiler. The requirement of complex equipment is a significant drawback of this process, and because of this, it is used only when stringent purity of the treated gas is required. Methanol solvents are licensed and manufactured by several companies, including Linde AG [11].

iii. N-Methyl-2-Pyrrolidone (NMP) (Purisol)

The parasol process uses NMP, licensed by Lurgi [11]. The process flow scheme is similar to the DEPG process but operates at ambient or very low temperatures of about 5 °F (– 15 °C) [18]. NMP has the highest selectivity for H_2S over CO_2 and high vapor pressure among all the physical solvents, which sometimes becomes the limitation of this process. This process is mainly suited to purifying high-pressure and high CO_2 concentration gas. This process does not require the removal of H_2S prior to the CO_2 absorption process. NMP has a relatively high vapor pressure as compared to DEPG or PC [4]. The regeneration of spent solvent is recommended by washing the treated gas and the off-gases with water. NMP cannot be used for gas dehydration if water wash is used.

iv. **Propylene Carbonate (PC) (Fluor)**

Fluor™ process is the most suitable physical absorption process for high CO_2 partial pressure gas streams [11]. Propylene carbonate is a physical solvent typically used when little or no H_2S is present in the gas mixture, and CO_2 removal is of prime importance. The FLUOR solvent has high CO_2 solubility, which enhances CO_2 loading. PC is less selective than DEPG and NMP for H_2S removal from gases containing CO_2. PC cannot be thermally regenerated since it becomes unstable at high temperatures of 200 °F or 93 °C) [14]. Moderate-to-high pressures ranging between 30 and 80 bar and ambient temperature are required for the Flour solvent's physical absorption process. Pre-removal of moisture and H_2S is required before the gas enters the absorption column. The spent solvent is flashed at low pressures to release CO_2 and recover the solvent. The lean solvent is recycled to the absorber, whereas highly pure CO_2 gas obtained during regeneration is compressed and stored. The high cost of Flour solvents is the principal disadvantage of this process [4].

6.3 Selection of Physical Solvents for Biogas Upgrading

The selection of a physical solvent for biogas upgrading depends on the process objectives, moisture in the feed gas, solvent cost, and availability. The solvent should have high selectivity for H_2S and CO_2 and thermal stability. The following table compares the essential properties of four physical solvents generally used. All these solvents are more selective for H_2S and CO_2 than CH_4 and H_2. Solubilities of various gases in solvents relative to carbon dioxide at 25 °C are also presented in Table 6.1.

6.4 Organic Scrubbing for Biogas Upgrading

Organic scrubbing is a well-established technology for CO_2 separation. Being a mature technology, it still occupies only a 6% market share in the biogas-upgrading sector [16]. It can achieve 98% biomethane quality with less than 2% CH_4 loss [1]. However, the value depends on the biogas composition, plant operating parameters, and optimization. The energy consumption is comparable to water scrubbing systems [3, 12]. Due to high CO_2 absorption capacity, CO_2 with high purity can be obtained as a by-product. Since moisture and other halogenated carbons can be absorbed in the physical solvent along with CO_2, this process has the advantage that it does not require their removal before the process.

The energy consumption for upgrading biogas absorption using organic solvents is comparable to water scrubbing. Energy is mainly consumed by the compressor, cooling machine, and feed pump. However, due to the high solubility of carbon dioxide in the organic solvents, the low flow rate of the organic solvents is required. Thus, the pump consumes less energy when pumping the organic solvent. The energy

Table 6.1 Characteristics of physical solvents [4, 11, 14, 18, 19]

Solvent	Process name	Viscosity (25 °C)	Specific gravity 25 °C (kg/m^3)	Vapor pressure at 25 °C (mm Hg)	CO_2 solubility (25 °C) (ft^3/U.S gal)	Relative solubilities of gases in physical solvents relative to CO_2 (25 °C)			Companies	Advantages	Disadvantages
						CH_4	CO_2	H_2S			
Selexol	Dimethyl ether of polyethylene glycol (DEPG)	5.8	1030	0.00073	0.485	0.066	1.0	8.82	Coastal Chemical Company (as Coastal AGR), Dow and UOP	• Easy regeneration by depressurization • Non-corrosive • Dehydrated gas leaves the absorber	• Elevated pressures have to be attained to achieve high efficiency
Rectisol	Methanol (MeOH)	0.6	785	125	0.425	0.051	1.0	7.06	Linde AG, Lurgi AG	• High chemical and thermal stability • Non-corrosive solvent • Non-foaming solvent	• High capital costs due to chilling of solvents • Low tempera-tures form amalgams • Complex process

(continued)

Table 6.1 (continued)

Solvent	Process name	Viscosity (25 °C)	Specific gravity 25 °C (kg/m^3)	Vapor pressure at 25 °C (mm Hg)	CO_2 solubility (25 °C) (ft^3/ U.S gal)	Relative solubilities of gases in physical solvents relative to CO_2 (25 °C)			Companies	Advantages	Disadvantages
						CH_4	CO_2	H_2S			
Purisol	N-Methyl-2-Pyrrolidone (NMP)	1.65	1027	0.40	0.477	0.072	1.0	10.2	Lurgi AG	• Non-foaming, non-corrosive solvent • High chemical and thermal stability • Low volatility	• High compression cost • Elevated pressures have to be attained to achieve high efficiency
Flour	Propylene carbonate (PC)	3.0	1195	0.085	0.455	0.038	1.0	3.29	Fluor Daniel, Inc	• Easy to use • Non-thermal regeneration • High solubility of CO_2 • Non-corrosive solvent	• High rates of solvent circulation • Expensive solvent

consumption depends upon the capacity of the plant. For a plant with a capacity of 200 Nm^3/h raw biogas, energy consumption is 0.275 kWh/Nm^3 which decreases with the plant's capacity. For a 1000 Nm^3/h capacity plant, energy consumption is 0.2–0.25 kWh Nm^3, which remains constant as the plant capacity increases. The process also requires heat (< 0.2kWh Nm^3 raw biogas) to regenerate the spent organic solvent.

This technology was developed in the early 2000s and has almost the exact investment costs. The capital costs decrease from 5000–4500 €/Nm^3/h for plants having a capacity of 250 Nm^3/h plants to 3500–2000 €/Nm^3/h for a scale of 1000 Nm^3/h [5]. Capital costs for large-scale plants having capacities over 1500 Nm^3/h are constant at 1500 €/Nm^3/h. Operating costs are mainly due to the electricity required for biogas and solvent pumping compression. For small-scale plants (100 m^3/h), typical operating costs are 13.8 ct/m^3 cleaned gas, which decreases to 9.0 ct/m^3 as the capacity increases to 500 m^3/h. Two to three percent of investment costs go toward maintenance [15]. Despite the fact that solvent regeneration often uses the waste heat from incineration or combustion facilities' exhaust gases, greater energy requirements in the range of 0.4–0.51 kWh/Nm^3 have been documented in the literature [5, 12]. High availability is essential for the plant's economy. However, it depends upon the age of the plant, operation load, and the number of hours it is operational in a day. Manufacturers guarantee an availability of 96–98% for this technology.

During physical absorption, raw biogas is initially compressed to 7–8 bar(a), then cooled and channeled from the bottom of the scrubbing column. The organic solvent is cooled and injected from the top of the column. The gas and the solvent interact counter currently inside the column. The column is filled randomly with packing material to increase the contact time and surface area between the solvent and the biogas. As the gas flows upward, more and more CO_2 gets absorbed into the organic solvent, and eventually, upgraded biogas is obtained from the top of the column. It is compressed to high pressures and dried before it is delivered to the gas grid or the fueling station [7].

In the regeneration process, the organic solvent leaving the absorption column's bottom is directed to a heat exchanger. It is heated to 40–80 °C and injected into a low-pressure flash column where the central part of the dissolved CH_4 and CO_2 is released. The recovered gases are recirculated back to the compressor. The spent solvent is regenerated by further heating it to higher temperatures. It enters the desorption/stripping column from the top, and air enters from the bottom. The stripping column is filled with random packing to increase the surface area. The solvent is regenerated at atmospheric pressures. The heat required in the process is exchanged as waste heat from the compressor—the process flow diagram for biogas upgrading is shown in Fig. 6.1 [8].

Physical solvents most commonly used in the biogas upgrading process are seloxol methanol [9]. These have significantly higher solubility of CO_2 than water. For example, Selexol® has five times higher CO_2 absorption capacity than water. This considerably decreases the solvent requirement for the process [17]. On the other hand, the high solubility of CO_2 in the solvent also makes the regeneration of organic solvents complex and energy-intensive. Similarly, due to the higher solubility of H_2S

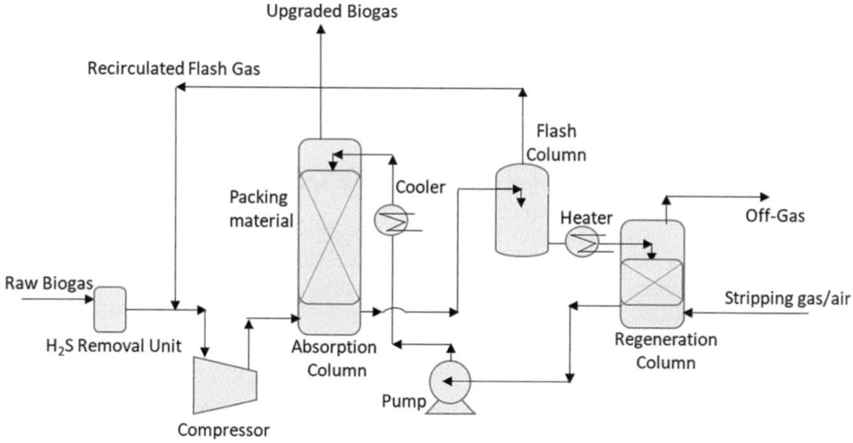

Fig. 6.1 Process flow diagram of biogas upgrading by organic scrubbing process. Reproduced with Permission from Publisher [8]

in Selexol® than CO_2, spent solvent is regenerated energy-intensive and carried out at high temperatures with the help of steam or inert gas to avoid a sulfur-mediated solvent deterioration. Therefore, to avoid high energy consumption, pre-removal of H_2S is recommended [6].

6.5 Conclusion

Physical absorption is one of the technologies developed for CO_2 separation. However, many parameters can still be improved to enhance this technology further so that it is feasible and economically used in the biogas upgrading industry to remove CO_2 from biogas. Future research should focus on developing a better solvent with high CO_2 absorption characteristics that can reduce operating costs and minimize equipment sizing. The spent solvent regeneration should be an effective and low-cost method with less energy requirement. Process equipment should be designed and configured to proficiently reuse waste heat and help meet the objectives of a better physical absorption and desorption process.

References

1. Angelidaki I, Treu L, Tsapekos P, Luo G, Campanaro S, Wenzel H, Kougias PG (2018) Biogas upgrading and utilization: current status and perspectives. Biotechnol Adv 36(2):452–466. https://doi.org/10.1016/j.biotechadv.2018.01.011
2. Ban ZH, Keong LK, Shariff AM (2014) Physical absorption of CO_2 capture: a review. Adv Mater Res 917:134–143. https://doi.org/10.4028/www.scientific.net/AMR.917.134

3. Bauer F, Hulteberg CP (2013) Biogas upgrading-review of commercial technologies. https://www.researchgate.net/publication/235932569
4. Burr B, Lyddon L (2008) A comparison of physical solvents for acid gas removal
5. Cucchiella F, D'Adamo I, Gastaldi M (2017) Sustainable waste management: waste to energy plant as an alternative to landfill. Energy Convers Manage 131:18–31. https://doi.org/10.1016/j.enconman.2016.11.012
6. de Hullu J, Maassen JIW, van Meel PA, Shazad S, Vaessen JMP (2008) Comparing different biogas upgrading techniques final report
7. Francisco López A, Lago Rodríguez T, Faraji Abdolmaleki S, Galera Martínez M, Bello Bugallo PM (2024) From biogas to biomethane: an in-depth review of upgrading technologies that enhance sustainability and reduce greenhouse gas emissions. Appl Sci (Switzerland) 14(6). https://doi.org/10.3390/app14062342
8. Kapoor R, Ghosh P, Kumar M, Vijay VK (2019) Evaluation of biogas upgrading technologies and future perspectives: a review. Environ Sci Pollut Res 26(12):11631–11661. https://doi.org/10.1007/s11356-019-04767-1
9. Khan IU, Othman MHD, Hashim H, Matsuura T, Ismail AF, Rezaei-DashtArzhandi M, Azlee IW (2017) Biogas as a renewable energy fuel—a review of biogas upgrading, utilisation and storage. Energy Convers Manag 150:277–294. https://doi.org/10.1016/j.enconman.2017.08.035
10. Mohd Pauzi MM, Azmi N, Lau KK (2022) Emerging solvent regeneration technologies for CO_2 capture through offshore natural gas purification processes. Sustainability (Switzerland) 14(7). https://doi.org/10.3390/su14074350
11. Mumford KA, Wu Y, Smith KH, Stevens GW (2015) Review of solvent based carbon-dioxide capture technologies. Front Chem Sci Eng 9(2):125–141. https://doi.org/10.1007/s11705-015-1514-6
12. Muñoz R, Meier L, Diaz I, Jeison D (2015) A review on the state-of-the-art of physical/chemical and biological technologies for biogas upgrading. Rev Environ Sci Biotechnol 14(4):727–759. https://doi.org/10.1007/s11157-015-9379-1
13. Ochedi FO, Yu J, Yu H, Liu Y, Hussain A (2021) Carbon dioxide capture using liquid absorption methods: a review. Environ Chem Lett 19(1):77–109. https://doi.org/10.1007/s10311-020-01093-8
14. Olajire AA (2010) CO_2 capture and separation technologies for end-of-pipe applications—a review. Energy 35(6):2610–2628. https://doi.org/10.1016/j.energy.2010.02.030
15. Patterson T, Esteves S, Dinsdale R, Guwy A (2011) An evaluation of the policy and techno-economic factors affecting the potential for biogas upgrading for transport fuel use in the UK. Energy Policy 39(3):1806–1816. https://doi.org/10.1016/j.enpol.2011.01.017
16. Thrän D, Krautz A, Scheftelowitz M, Lenz V, Liebetrau J, Daniel-Gromke J, Zeymer M, Nelles M (2014) Auswirkungen der gegenwärtig diskutierten Novellierungs-vorschläge für das EEG-2014. www.dbfz.de
17. Tock JY, Lai CL, Lee KT, Tan KT, Bhatia S (2010) Banana biomass as potential renewable energy resource: a Malaysian case study. Renew Sustain Energy Rev 14(2):798–805. https://doi.org/10.1016/j.rser.2009.10.010
18. Vega F, Cano M, Camino S, Fernández LMG, Portillo E, Navarrete B (2018) Solvents for carbon dioxide capture. In: Carbon dioxide chemistry, capture and oil recovery. InTech. https://doi.org/10.5772/intechopen.71443
19. Xu Y, Schutte RP, Hepler LG (1992) Solubilities of carbon dioxide, hydrogen sulfide and sulfur dioxide in physical solvents. Can J Chem Eng 70(3):569–573

Chapter 7
Chemical Scrubbing Technology

7.1 Introduction to Chemical Scrubbing

Chemical absorption is an exothermic process of forming reversible solid chemical bonds between the solute and the solvent. Initially, the solute gets absorbed in the solvent-based on physical absorption, followed by a chemical reaction between the two phases. This strengthens the bond between the unwanted gas component (solute) and the scrubbing liquid (solvent). The endothermic stripping process regenerates the solvent at high temperatures and low pressures. Chemical absorption techniques are especially useful in situations when the partial pressure of carbon dioxide (CO_2) is low and the leftover gas requires little acid gas.

7.2 Chemical Solvents for the Absorption Process

Solvents are liquids in which the solute dissolves to form a solution. Different characteristics of the solvents affect the efficiency of gas absorption. Thus, solvent selection becomes a vital task for the process. The solvent choice is based on the solubility, molecular weight, and vapor pressure of the solute, foaming, corrosion, reaction kinetics, regeneration energy requirement, and cyclic capacity of the solvent. Cost and environmental contamination also affect solvent selection [23]. Therefore, it is critical to select an appropriate solvent for the proficient and cost-effective performance of the absorber with the qualities described below:

(i) **Solubility**: The absorption process is based on the solubility of the solute in the solvent. Hence, a judiciously selected solvent in which the solute is selectively soluble is generally favored. The solubility of the solute gas should be high in the solvent to increase the absorption and loading capacity rate and decrease the quantity of solvent required.

V. K. Vijay et al., *Compressed Biogas (CBG)*, Green Energy and Technology,
https://doi.org/10.1007/978-981-96-5697-4_7

(ii) **Volatility**: The property of the solvent which describes how quickly a substance vaporizes. The solvent should possess low volatility or vapor pressure to minimize solvent losses due to vaporization.
(iii) **Viscosity:** The property of the solvent describing its ability to flow quickly and effortlessly. Solvents with low viscosity are preferred for high absorption rates and low-pressure drops. If a solvent with high viscosity is used, it results in flooding even at low liquid and gas flow rates.
(iv) **Corrosiveness**: Using a less corrosive solvent decreases the maintenance cost and the cost of anti-corrosive coating on the material.
(v) **Cost**: The solvent used for the absorption process should not be expensive to combat the vaporization and other solvent losses.
(vi) **Miscellaneous**: A solvent should be non-hazardous, non-toxic, non-foaming, and non-inflammable.

When selecting a solvent, a balance between the properties, efficiency, cost, and environmental effects should be considered [17].

7.2.1 Amines

Amines have been widely used in gas absorption for 70 years, primarily in the oil and chemical industries, to remove acidic H_2S and CO_2 from gas streams. Amines are organic compounds that are created by substituting an alkyl or aryl group with one or more hydrogen atoms in ammonia. Since alkanol amines include at least one hydroxyl group, their solubility in water is increased, and their vapor pressure is lowered. The necessary alkalinity to absorb acidic gasses in aqueous solutions is provided by one amino group. Monoethanolamine (MEA), diglycolamine (DGA), diethanolamine (DEA), triethanolamine (TEA), methyl diethanolamine (MDEA), 2-amino-2methyl-1propanol (AMP), and piperazine (PZ) are the most often utilized amines for the removal of CO_2 and H_2S [1].

The primary criteria used to choose amines are absorption capacity, acid gas partial pressures, kinetics of the absorption reaction, regenerative potential, and the necessary purity of the treated gas. Acid gas partial pressures ranging from low to medium are often suitable for amine-based purification procedures, which lead to lower process costs as acid gas partial pressures rise. Different amine concentrations are used depending on the type of amine used and what gases must be removed. For example, 20% monoethanolamine is used to remove H_2S and CO_2 and about 32% for removing only CO_2 [1].

The choice of amine depends upon several factors and may be pretty subjective. However, no ideal solution is available for optimum operating conditions and high acid gas separation. Amines are selected based on experimental data, operating experiences, and economic analysis and are available for various conditions. Amines are subcategorized as primary, secondary, and tertiary based on the degree of hydrocarbon substitution as follows:

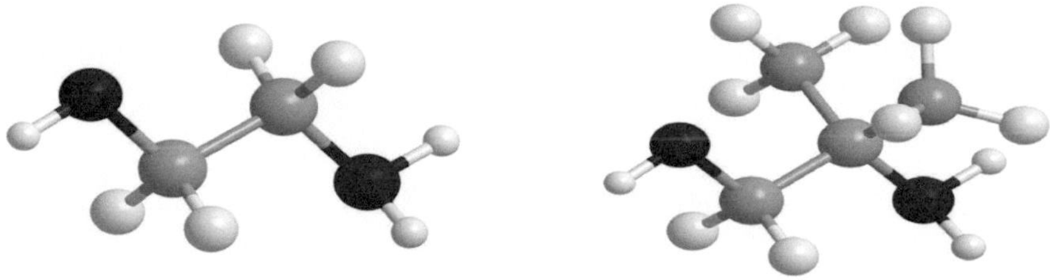

Fig. 7.1 Chemical structure of monoethanolamine (MEA). Reproduced with Permission from Elsevier [28]

a. **Primary Amines**—An alkyl or aryl group replaces one hydrogen atom of NH3 in primary amines. Two types of primary amines are most commonly used in the gas separation industry: monoethanolamine (MEA) and 2-(2-aminoethoxy) ethanol (DGA), as shown in Fig. 7.1 , respectively.

   - *Monoethanolamine (MEA)*

Aqueous MEA is popularly used to remove H_2S and CO_2 from gas mixtures. It is preferred for efficient removal of low concentrations of H_2S and CO_2 at low pressures. MEA has a high CO_2 absorption capacity (0.55 kg CO_2 kg MEA-1) [19]. MEA has low molecular weight and high alkalinity, resulting in high capacity at moderate concentrations. However, MEA is corrosive mainly if the amine concentration is more than 20% and the solutions are loaded with high concentrations of acidic gases. A significant drawback of MEA is its high heat of reaction with CO_2 and H_2S. This results in high energy consumption during the regeneration of spent MEA [1]. MEA also possesses high vapor pressure, which causes significant vaporization losses during absorption processes.

- *Diglycolamine (DGA)*

DGA (2-(2-aminoethoxy) ethanol) is similar to MEA in many respects except for the fact that DGA solvent has low vapor pressure, which makes it applicable for gases having relatively high concentrations of acidic gases like CO_2 and H_2S (40–60 wt%), which results in low circulation rates [13].

b. **Secondary amines** have two organic substituents (alkyl, aryl, or both) bound to N and one hydrogen (or no hydrogen if one of the substituent bonds is double). Diethanolamine (DEA), as shown in Fig. 7.2, is a popular secondary amine used for CO_2 and H_2S removal.

   - *Diethanolamine (DEA)*

DEAhas been used for many years to remove CO_2 and H_2S for gas mixtures. DEA has a higher acid gas loading capacity than MEA. DEA is appropriate for low-pressure operations. DEA has the advantage of low heat of reaction, which

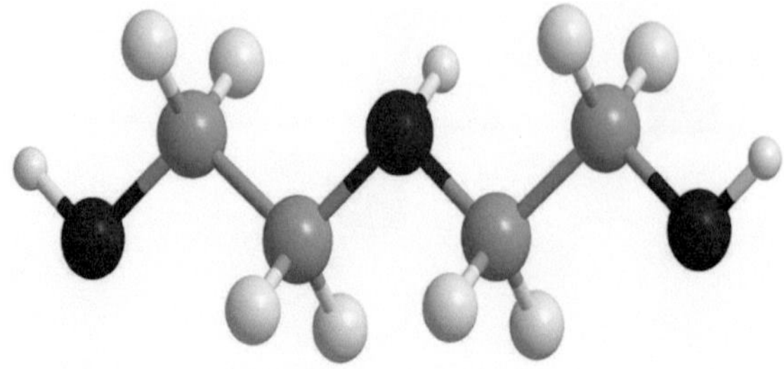

Fig. 7.2 Chemical structure of diethanolamine (DEA). Reproduced with Permission from Elsevier [28]

requires lower energy for regeneration, but regeneration is more complex due to the requirement of vacuum distillation [13].

c. **Tertiary Amines**—All three hydrogen atoms are swapped out for organic substituents in tertiary amines. Triethanolamine (TEA) and methyldiethanolamine (MDEA) are tertiary amines in the gas purification industry. However, because of its higher equivalent weight, low absorption capacity, reactivity, and relatively poor stability, TEA is not so famous for separating acidic gases from gas mixtures. The most popular amine in this group is N-methyl diethanolamine (MDEA), as shown in Fig. 7.3. [13].

- *Methyldiethanolamine (MDEA)*

MDEA is the most preferred nonselective amine used to separate high concentrations of CO_2 and H_2S from gas mixtures. It is the most chosen amine because of its high absorption capacity, low regeneration energy demand, tendency to generate

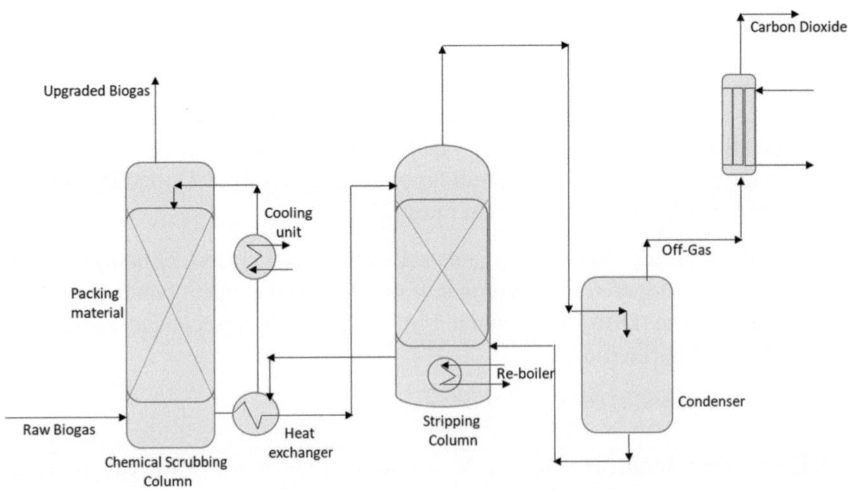

Fig. 7.3 Biogas upgrading by chemical absorption (amine scrubbing). Reproduced with Permission from Publisher [11]

degradation products, anti-corrosion, and exceptional thermal and chemical stability. Its absorption capacity is poor, and its kinetic rate is quite slow. MDEA can be employed in high concentrations of up to 60% weight percent due to its low vapor pressure and basicity, which leads to low circulation rates, compact processing units, and costs.

ii. **Mixed Amines**

Blended amines are amalgams of alkanol amines, i.e., a mixture of two or more amines in different concentrations and compositions. Blended amines syndicate the preferred characteristics of different absorbents while suppressing their unwanted characteristics. Blended amines integrate high reaction kinetics of the primary or secondary amines with low heat of reaction of the tertiary amine. The resulting tertiary amine has high reaction kinetics and lower heat of regeneration. The mixture of primary amines, such as MEA, or secondary amines, like DEA, with a tertiary amine, such as MDEA, increases the rate of CO_2 absorption considerably [13]. Blended amine solutions achieve the desired degree of separation of CO_2 from the gas mixture, thus reducing capital and operating costs [22]. The absorption capacities and regeneration energy requirements are highest for MEA, followed by DEA and MDEA, while blended alkanol amines are between their original alkanol amines.

iii. **Sterically Hindered Amines**

A primary amine is joined to a tertiary carbon atom, or a secondary or tertiary carbon atom is joined to a secondary or tertiary carbon atom in sterically hindered amines. For CO_2 absorption, the most often used sterically hindered amine is 2-amino-2-methyl-1-propanol (AMP). AMP is comparatively slower in oxidative degradation and thermally stronger than MEA [25]. The large tertiary carbon near the amino functional group creates an unstable carbamate, resulting in a faster reaction with CO_2 than tertiary amines while lowering solvent regeneration costs.

iv. **Amine Activators**

Activated amine solutions increase the absorption rates by doping conventional amine with a small quantity of an accelerator (activator). Piperazine (PZ), a diamine, is a popular activator for improving absorption kinetics and increasing solvent capacity, such as doped MEA/PZ and MDEA/PZ blends. Piperazine is a promising solvent due to its high loading capacity (two moles of CO_2 per one mole of PZ), high reaction rate with CO_2 (more significant than that of MEA, DEA, MDEA, and AMP), and resistance to thermal and oxidative degradation [1].

7.2.2 *Alkali Solutions*

The second most popular solvents used for the chemical absorption of CO_2 from a gas mixture are solutions of alkali salts like sodium hydroxide, potassium hydroxide, and calcium hydroxide. These solvents have been used for over 50 years in the acid gas removal industry. Sodium hydroxide, a robust alkaline solution, is the most frequently used caustic solvent for CO_2 absorption. NaOH has the advantage of being more abundant, readily available, and less expensive than MEA. Theoretically, it has a high CO_2 capture capacity [22].

Absorption using alkali salts (NaOH and KOH) results in the formation of thermally stable spent solvents ($NaHCO_3$ and K_2CO_3). Regeneration of these solutions is complex and challenging since it requires a high temperature of 160 °C for the decomposition of $NaHCO_3$ into Na_2CO_3, H_2O, and CO_2 and an even higher temperature of 800 °C to form a stable form of Na_2O [1]. However, this process is highly inefficient and produces low alkalinity solvent. Therefore, the regeneration of alkali solutions is complex, expensive, and highly energy-demanding.

7.2.3 Ammonia

Ammonia is a promising alternative to organic chemicals. It is non-degradable and non-corrosive, and due to its weaker bonding with CO_2, it requires up to 75% less regeneration energy than MEA. Moreover, aqueous ammonia has a 1.76 $kgCO_2$/kg NH3 absorption capacity, three times higher than MEA and only one-sixth of the cost [1, 22]. A significant drawback of using ammonia for CO_2 absorption from gas mixtures is its high-saturated vapor pressure, which results in its introduction in the gas phase. This necessitates installing ammonia-cleaning machinery near the gas absorption column's output, which will raise the energy and capital costs.

7.3 Solvents for Biogas Upgrading

Amines, which are derivatives of ammonia (NH3), are frequently employed as chemical solvents for the removal of acidic gases such as CO_2 and H_2S. According to [16], amines such as mono, di, and triethanolamine are commonly utilized in the upgrading of biogas. These amines include, but are not limited to, monoethanolamine (MEA), diglycolamine (DGA), diethanolamine (DEA), triethanolamine (TEA), methyldiethanolamine (MDEA), and sterically hindered amines like 2-amino-2-methyl-1-propanol (AMP) and piperazine (PZ).

The least expensive amine for used as a cleaning agent to absorb CO_2 is MEA [1]. It is the most widely utilized amine because of its high absorption capabilities and selectivity for CO_2 [19]. Up to 30 weight percent MEA has been effectively used in the commercial sector with CO_2 removal efficiencies of 80–95%. MDEA solvents can be employed at high concentrations of up to 60% weight percent because to their low vapor pressure and basicity, which leads to smaller plants, lower circulation rates, and reduced costs [6].

MEA > DEA > AMP > MDEA is the order in which the CO_2 absorption rates of MEA, DEA, and AMP are higher than MDEA, while the order in which the CO_2 absorption capacity is higher than MDEA is AMP > MEA > DEA > MDEA [25]. Activated MDEA (AMDEA), a combination of MDEA and piperazine (PZ), is gaining popularity for CO_2 absorption from biogas due to its higher absorption capacity than MDEA [21]. This is because MDEA contains a tertiary amine that increases the reaction rate for CO_2 absorption, while PZ contains primary and secondary amines [15]. An alternative chemical absorbent for removing CO_2 from biogas is aqueous-free ammonia (NH3) [20]. Because of the weaker interaction between CO_2 and ammonia, it has the largest absorption capacity, is non-degradable, and takes up to 75% less energy for regeneration than MEA [18, 19].

Aqueous solutions of alkaline salts such calcium hydroxides, sodium hydroxides, potassium hydroxides, and ammonium hydroxides are commonly utilized as inorganic solvents for chemical scrubbing [1]. Since CO_2 dissolves more readily in sodium hydroxide than in amines, 1.39 tons of monoethanolamine should be needed to absorb 1 ton of CO_2 instead of 0.9 tons of sodium hydroxide [3]. Alkali solutions are expensive due to the high-energy requirement (Table 7.1).

7.4 Chemical Scrubbing of Biogas Upgrading

The configuration of the chemical scrubbing system is simple and highly efficient because of the use of highly reactive chemical absorbents [2]. The process consists of a packed bed scrubbing column, a reboiler, and a desorption unit. Initially, raw biogas compressed at 1–2 bars is injected from the bottom of the scrubbing column [5]. The column is filled with packing material. Either structured or random packings can be used in the chemical scrubbing process, as Chemical scrubbing of biogas is an exothermic reaction. The high pH of the chemical solvent controls the possibility of packing blockage caused by biomass development [27]. Throughout the process, the absorber temperature rises from 20 to 65 °C [12]. The production of intermediate chemical molecules (CO32–, HCO3–) by the exothermic process leads to an accelerated mass transfer of CO_2 from biogas, hence increasing the capacity for CO_2 absorption. As a result, the units become more compact and smaller and have lower rates of absorbent recycling. Via a heat exchanger, the wasted solution, which is rich in CO_2 and H_2S, is pushed to a stripping column from the absorber's bottom. At a pressure of 1.5 bars, the wasted is heated to 120–160 °C in the stripping column for regeneration [12, 24]. A reboiler is installed in the stripping column to supply reaction heat for the desorption of CO_2 from the spent chemical solution. The chemical bonds created during absorption are broken by the heat. To release the CO_2 that has been held in the exhaust, the steam is cooled in a condenser before being directed to the stripper. Biogas upgrading by chemical absorption is shown in Fig. 7.3.

With a CH_4 loss of less than 0.1%, biomethane with a high concentration of > 99% CH_4 exits the column from the top. Although CH_4 loss of up to 4% has been reported in some modeling reports due to CH_4 dissolution in alkanol amine [4, 9,

Table 7.1 Absorption capacity, advantages, and disadvantages of some chemical absorbents [1, 17, 26]

Name	Absorption capacity ($molCO_2$/mol amine)	Advantages	Disadvantages
Monoethanolamine (MEA)	0.45–52 [26]	• Elevated responsiveness • Minimal solvent expense • Reasonable thermal stability, low molecular weight, and degradation rate	• High energy need for regeneration • Inefficient method • Significant recirculation rates are caused by relatively poor CO_2 loading capacity, which ultimately necessitates huge equipment sizes and expensive capital costs • High vapor pressure, high solvent degradation, and losses
Diethanolamine (DEA)	0.21–0.81 [26]	• Reduced energy needed for regeneration • Low reaction heat	• High degradation and losses, but not as much as with MEA • The process of renewing the wasted DEA solutions is more intricate since vacuum distillation can be necessary
N-methyldiethanolamine (MDEA)	0.20–0.81 [26]	• Relatively low regeneration energy requirement for regeneration • Less degradable and corrosive • Low vapor pressure and basicity • Lower circulation rates, smaller plant size, and costs	• Relatively slow kinetics • Low absorption capacity

(continued)

Table 7.1 (continued)

Name	Absorption capacity ($molCO_2$/mol amine)	Advantages	Disadvantages
2-amino-2-methyl-1-propanol (AMP)	0.84	• Degradation of oxidation is slow • Greater resistance to heat deterioration than MEA • AMP has a higher capacity to absorb CO_2 than other amines • Quicker than with tertiary amines in interaction with CO_2 • Inexpensive solvent regeneration	• Absorption rate is low as compared to MDEA
MDEA and piperazine (PZ), known as activated MDEA (AMDEA)	0.41 (Rochelle et al.)	• High absorption capacity significantly as compared to MDEA	● Limited solvent solubility as solid precipitation occurs at both low and high CO_2 loading
Ammonia (NH3)	1.76 [19]	• Non-degradable • Non-corrosive	• Due to its extreme volatility, NH3 poses a risk of air pollution • Ammonia with a high-saturated vapor pressure introduces NH3 into the gas phase • However, ammonia washing equipment at the gas absorption column's output will raise energy consumption and capital costs

14, 29], CH_4 recovery of 99.5–99.9% can be attained with high purities due to the low solubility of CH_4 in chemical solvents [10, 12]. This method has the benefit of tolerating up to 300 ppmv of H_2S found in biogas, which can be completely absorbed in the amine scrubber [21]. However, in order to avoid solvent toxicity, it is strongly advised to remove excessive quantities of H_2S before scrubbing.

The biomethane produced via chemical scrubbing-based biogas upgrading has a high concentration of more than 99%. Because CH_4 is not soluble in chemical solvents, the procedure offers the advantage of a high CH_4 recovery of 99.5–99.9%, which is easily achieved [10, 12]. Although this technique may withstand H_2S concentrations of up to 300 ppmv, it is strongly advised to remove high H_2S concentrations prior to scrubbing in order to avoid solvent poisoning [11]. Like other upgrading technologies, chemical scrubbing does not separate nitrogen from biogas, and the oxygen irreversibly damages the scrubber as it oxidizes amines and hinders CO_2 absorption.

The chemical scrubbing process requires significant energy. Electricity consumption is marginal in the 0.1–0.25 KWh/Nm^3 raw biogas range. It is mainly for low-pressure gas compression and solvent pumping. Energy is mainly required for heat during the regeneration of the chemical solvent. A typical value for the heat demand is 0.55–0.75 kWh/Nm^3, which is required in heating the stripper [11].

Because regeneration requires a high heat demand, energy consumption is quite significant even with its high CO_2 removal efficiency. Additionally, foaming, salt precipitation, and O_2 poisoning of amine and other compounds are possible outcomes. In comparison to other methods, this system is more difficult due to foaming, amine degradation, solvent losses and make-up issues, corrosion, and operating issues. Only a small portion of this developed technology—22% of the global upgrading market—is attributed to these drawbacks [30].

The investment costs for the chemical scrubbing process are slightly high. Small-scale plants (100 Nm^3/h) have high investment costs in the 9500 € /Nm^3/h range. As the capacities rise, the investment cost of chemical cleansing falls. The investment costs are 3500 €/Nm^3/h for a 500 Nm^3/h capacity and 1800 €/Nm^3/h for upgrading facilities with a 1500 Nm^3/h capacity [8]. The expenses of chemicals, anti-foaming agents, and chemical make-up are marginal, but the electricity needed for liquid pumping and gas compression is reasonable, and the energy needed for amine regeneration accounts for a large portion of the operating costs [5, 8, 21]

As treatment capacities rise, chemical scrubbing's investment cost falls. The investment costs for upgrading facilities with a capacity of 1800 Nm^3/h are 1500 € per Nm^3/h, whereas for a treatment capacity of 600 Nm^3/h, they are 3200 €/Nm^3/h [5, 8]. The energy needed for amine regeneration (0.55 kWh Nm^3) is the main source of operating costs, even though the costs of chemicals, anti-foaming agents, and chemical make-up are negligible. Gas compression and liquid pumping have reasonable electricity requirements in the range of 0.12–0.15 kWh Nm^3 [5, 8, 21].

7.5 Conclusion

Chemical absorption of CO_2 is a promising technology for biogas upgrading. This mature technology for CO_2 removal and capture from postcombustion gases is still being developed for biogas upgrading. Several plants are based on the chemical absorption of CO_2 using amines, and caustic salts, which can upgrade biogas to high-concentration biomethane. Amine scrubbing is the most advanced method for chemically upgrading biogas that is currently accessible. To counteract the deterioration of solvents and energy loss during large-scale biogas upgrading, more study is necessary. Future studies ought to concentrate on assessing and developing the viability of upgrading biogas by the use of mixed amines (primary, secondary, tertiary, and sterically inhibited). The caustic solvent scrubbing also requires more research and development to reduce the energy requirements for the regeneration of the spent solvent. Chemical scrubbing-based biogas upgrading systems should focus on optimizing and reducing the environmental impact of the process. Moreover, the conversion of spent chemicals to value-added chemicals should also be considered to increase the process's economy.

References

1. Abdeen FRH, Mel M, Jami MS, Ihsan SI, Ismail AF (2016) A review of chemical absorption of carbon dioxide for biogas upgrading. Chin J Chem Eng 24(6):693–702. https://doi.org/10.1016/j.cjche.2016.05.006
2. Andriani D, Wresta A, Atmaja TD, Saepudin A (2014) A review on optimization production and upgrading biogas through CO_2 removal using various techniques. Appl Biochem Biotechnol 172(4):1909–1928. https://doi.org/10.1007/s12010-013-0652-x
3. Angelidaki I, Treu L, Tsapekos P, Luo G, Campanaro S, Wenzel H, Kougias PG (2018) Biogas upgrading and utilization: current status and perspectives. Biotechnol Adv 36(2):452–466. https://doi.org/10.1016/j.biotechadv.2018.01.011
4. Awe OW, Zhao Y, Nzihou A, Minh DP, Lyczko N (2017) A review of biogas utilisation, purification and upgrading technologies. Waste Biomass Valorization 8(2):267–283. https://doi.org/10.1007/s12649-016-9826-4
5. Bauer F, Persson T, Hulteberg C, Tamm D (2013) Biogas upgrading—technology overview, comparison and perspectives for the future. Biofuels Bioprod Bioref 7:499–511. https://doi.org/10.1002/bbb.1423
6. Bernhardsen IM, Knuutila HK (2017) A review of potential amine solvents for CO_2 absorption process: absorption capacity, cyclic capacity and pKa. Int J Greenh Gas Control 61:27–48. https://doi.org/10.1016/j.ijggc.2017.03.021
7. Budzianowski WM, Wylock CE, Marciniak PA (2017) Power requirements of biogas upgrading by water scrubbing and biomethane compression: comparative analysis of various plant configurations. Energy Convers Manage 141:2–19. https://doi.org/10.1016/j.enconman.2016.03.018
8. Cucchiella F, D'Adamo I, Gastaldi M (2017) Sustainable waste management: waste to energy plant as an alternative to landfill. Energy Convers Manage 131:18–31. https://doi.org/10.1016/j.enconman.2016.11.012
9. Huertas JI, Giraldo N, Izquierdo S (2011) Removal of H_2S and CO_2 from biogas by amine absorption. www.intechopen.com
10. Kadam R, Panwar NL (2017) Recent advancement in biogas enrichment and its applications. Renew Sustain Energy Rev 73:892–903. https://doi.org/10.1016/j.rser.2017.01.167
11. Kapoor R, Ghosh P, Kumar M, Vijay VK (2019) Evaluation of biogas upgrading technologies and future perspectives: a review. Environ Sci Pollut Res 26(12):11631–11661). https://doi.org/10.1007/s11356-019-04767-1
12. Khan IU, Othman MHD, Hashim H, Matsuura T, Ismail AF, Rezaei-DashtArzhandi M, Azlee IW (2017) Biogas as a renewable energy fuel—a review of biogas upgrading, utilisation

and storage. Energy Convers Manag 150:277–294. https://doi.org/10.1016/j.enconman.2017.08.035
13. Kohl AL, Nielsen RB (1997) Gas purification (5th ed.). Gulf Publishing Company
14. Lasocki J, Kołodziejczyk K, Matuszewska A (2015) Laboratory-scale investigation of biogas treatment by removal of hydrogen sulfide and carbon dioxide. Pol J Environ Stud 24(3):1427–1434. https://doi.org/10.15244/pjoes/35283
15. Li Y, Yi H, Tang X, Li F, Yuan Q (2013) Adsorption separation of CO_2/CH_4 gas mixture on the commercial zeolites at atmospheric pressure. Chem Eng J 229:50–56. https://doi.org/10.1016/j.cej.2013.05.101
16. Maile I, Muzenda E (2014) A review of biogas purification through chemical absorption. https://doi.org/10.15242/iie.e1114022
17. Maile OI, Tesfagiorgis HB, Muzenda E (2015) Factors influencing chemical absorption of CO_2 and H_2S in biogas purification: a review. Newswood Limited, International Association of Engineers
18. Makhloufi C, Lasseuguette E, Remigy JC, Belaissaoui B, Roizard D, Favre E (2014) Ammonia based CO_2 capture process using hollow fiber membrane contactors. J Membr Sci 455:236–246. https://doi.org/10.1016/j.memsci.2013.12.063
19. Mani S, Sokhansanj S, Bi X, Turhollow A (2006) Economics of producing fuel pellets from biomass. Appl Eng Agric 22(3):421–426. https://doi.org/10.13031/2013.20447
20. McLeod A, Jefferson B, McAdam EJ (2014) Biogas upgrading by chemical absorption using ammonia rich absorbents derived from wastewater. Water Res 67:175–186. https://doi.org/10.1016/j.watres.2014.09.010
21. Muñoz R, Meier L, Diaz I, Jeison D (2015) A review on the state-of-the-art of physical/chemical and biological technologies for biogas upgrading. Rev Environ Sci Biotechnol 14(4):727–759. https://doi.org/10.1007/s11157-015-9379-1
22. Olajire AA (2010) CO_2 capture and separation technologies for end-of-pipe applications—a review. Energy 35(6):2610–2628. https://doi.org/10.1016/j.energy.2010.02.030
23. Pellegrini LA, Gilardi M, Giudici F, Spatolisano E (2021) New solvents for CO_2 and H_{2S} removal from gaseous streams. Energies 14(20). https://doi.org/10.3390/en14206687
24. Petersson A, Wellinger A (2009) Biogas upgrading technologies-developments and innovations. www.iea-biogas.net
25. Rinprasertmeechai S, Chavadej S, Rangsunvigit P, Kulprathipanja S (2012) Carbon dioxide removal from flue gas using amine-based hybrid solvent absorption
26. Rufford TE, Smart S, Watson GCY, Graham BF, Boxall J, Diniz da Costa JC, May EF (2012) The removal of CO_2 and N_2 from natural gas: a review of conventional and emerging process technologies. J Pet Sci Eng 94–95:123–154.https://doi.org/10.1016/j.petrol.2012.06.016
27. Ryckebosch E, Drouillon M, Vervaeren H (2011) Techniques for transformation of biogas to biomethane. Biomass Bioenergy 35(5):1633–1645. https://doi.org/10.1016/j.biombioe.2011.02.033
28. Shokrollahi F, Lau KK, Partoon B, Smith AM (2022) A review on the selection criteria for slow and medium kinetic solvents used in CO_2 absorption for natural gas purification. J Nat Gas Sci Eng 98https://doi.org/10.1016/j.jngse.2021.104390
29. Sun Q, Li H, Yan J, Liu L, Yu Z, Yu X (2015) Selection of appropriate biogas upgrading technology-a review of biogas cleaning, upgrading and utilisation. Renew Sustain Energy Rev 51:521–532. https://doi.org/10.1016/J.RSER.2015.06.029
30. Thrän D, Krautz A, Scheftelowitz M, Lenz V, Liebetrau J, Daniel-Gromke J, Zeymer M, Nelles M (2014) Auswirkungen der gegenwärtig diskutierten Novellierungs-vorschläge für das EEG-2014. www.dbfz.de
31. Vega F, Cano M, Camino S, Fernández LMG, Portillo E, Navarrete B (2018) Solvents for carbon dioxide capture. In: Carbon dioxide chemistry, capture and oil recovery. InTech. https://doi.org/10.5772/intechopen.71443

Chapter 8
Membrane Separation

8.1 Introduction

In recent years, membrane technology has gained immense importance in the chemical industry, addressing various applications since the 1960s. It is a promising and competitive alternative to the commercially available and conventional gas absorption technologies. Nowadays, numerous plants are using this technique to separate gases. The membrane process has been applied for liquid treatment for many years, using reverse osmosis, ultrafiltration, and microfiltration. In recent years, membranes have also been applied to natural gas processing. Therefore, membrane technology can be easily adapted to upgrade biogas. The membrane-based biogas upgrade method is an efficient and easy-to-scale-up technology among all the upgrading technologies.

8.1.1 Mechanism of Membrane Separation

Membrane-based gas separation is based upon two main parameters: selective permeability and the preferential separation of a gaseous constituent from a mixture of gases through a semi-permeable membrane [5].

The membrane, the pivotal component of the unit, serves as a barrier, allowing only specific gases to permeate while retaining other constituents in the effluent gas stream. Its selectivity, where some constituents can pass through the membrane wall more rapidly than others, is a critical factor in the separation process. The degree of separation is directly linked to the flow rate of different molecules between two gaseous phases, a liquid and a gas phase, across the membrane wall [18]. The membrane acts as an obstruction for the hydrodynamic flow of the usually miscible fluid phases. The diffusion rate of the molecules is influenced by the partial pressure, the thickness of the membrane wall, and the chemical solubility of the substance.

V. K. Vijay et al., *Compressed Biogas (CBG)*, Green Energy and Technology,
https://doi.org/10.1007/978-981-96-5697-4_8

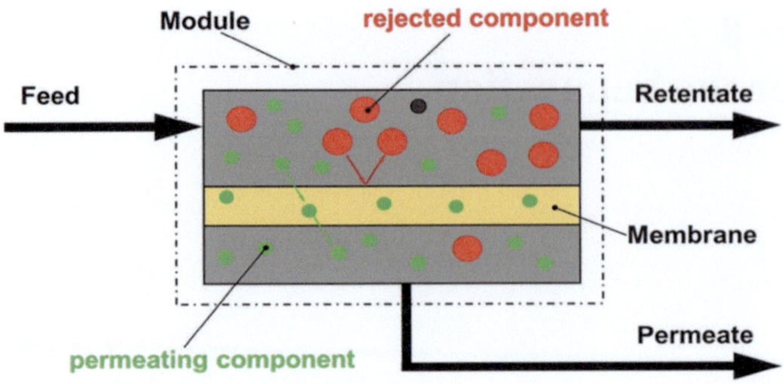

Fig. 8.1 Basic membrane separation process. *Reproduced under license CC-BY-4.0* [1]

Permeate and retentate are the two primary gas streams that make up the input gas stream. The permeate is the portion of the incoming gas stream that diffuses or passes through the semi-permeable membrane, while the retentate is the remaining portion that does not. This concept is illustrated in Fig. 8.1 [27].

8.1.2 Membrane Classification

A wide variety of membranes are available for the separation process, offering numerous options for innovation and improvement. An efficient membrane-based gas separation technique requires a membrane material with high-performance capacity [35]. The following characteristics of the membrane material are required to be examined to accomplish a membrane process with high proficiency:

- Transport properties of the membrane-like permeance and selectivity.
- Mechanical, chemical, and thermal strength at specific operating conditions.
- Sturdiness of membrane material over a long period.
- Modular design of the membrane unit.
- Optimization and integration of the designed process.

Numerous membranes have been developed and are commercially available with specific characteristics for separating CO_2 and CH_4 from a gas mixture [26]. For instance, polymeric membranes like cellulose acetate and polyimide are commonly used. Typically, membranes can be classified as porous, non-porous, and carrier membranes.

8.1.3 Porous Membranes

Porous membranes are based on the principle of size rejection. Four diffusion mechanisms can be exploited to separate gases in porous membranes: Knudsen diffusion (molecular weight), surface diffusion, capillary condensation, and molecular sieving. The difference in pressure, concentration, or electrical potential gradient of the gaseous constituents across the membrane results in diffusion from one side to the other of the membrane. They are often made of zeolites, glass, metal, or polymer. Membranes of this type are seldom used in gas separation. However, they are generally applied for liquid separations, e.g., membranes with pore size from 0.1 to 10 μm are used in microfiltration and 0.001–0.1 μm for ultrafiltration. Single modules with porous membranes have low efficiency; therefore, multiple separation stages are required for high-scale purification [17].

8.1.4 Non-porous (Dense) Membranes

Dense membranes dominate the gas permeation market. Non-porous membranes are dense polymers that depend on the solution-diffusion mechanism to separate gases. In this process, permeation is based on the solubility and diffusivity of the gas; at higher values, more permeate can pass through the membrane wall. According to the solution-diffusion mechanism, first, the gaseous molecule is adsorbed on the surface of the inlet portion of the membrane; after that, it absorbs and diffuses in the membrane, and finally, it desorbs from the other side, and permeate can be obtained [34]. Due to the difference in solution and diffusion coefficient of gases, separation is possible. Both effects are considered for gas species j with permeability P.

Kinetic and thermodynamic characteristics govern permeability and mechanism of separation. The size (diameter) of the gas molecules under consideration determines the kinetic parameter, also known as the diffusivity coefficient (D), which measures the flow of molecules through the membrane's polymer chains. As the gas's kinetic diameter increases, the diffusion coefficient falls. The solubility coefficient (S) is another name for the thermodynamic parameter [18]. As demonstrated in Eq. 8.1, permeability "P," which is the product of diffusivity and solubility coefficient, denotes the molecules' capacity to flow through a semi-permeable membrane wall.

$$P_j = D_j S_j. \tag{8.1}$$

Permeability only reveals the material properties, and the membrane's thickness is not considered. Selectivity is defined based on the slowest permeating species within a gas mixture. A widely used unit of permeability "P" is 1 Barrer = 10^{-10} cm^3 (STP) cm^{-2} s^{-1} cmHg^{-1} [2].

Selectivity "α" is the other parameter signifying the separation performance of the membrane [7]. The polymer's molecular structure controls it, and the membrane can separate a mixture of two gaseous constituents (*A* and *B*). The ideal selectivity of a membrane, as shown in Eq. 8.2, is given by the ratio of the permeability coefficients of two different gas species, which in turn comprises a diffusivity term (D_{CO_2}/D_{CH_4}) and a solubility term (S_{CO_2}/S_{CH_4}). This ratio is often called perm-selectivity or selectivity (α_{AB}).

$$\alpha \text{ ideal} = \frac{P_{CO_2}}{P_{CH_4}} = \frac{D_{CO_2}}{D_{CH_4}} * \frac{S_{CO_2}}{S_{CH_4}}. \tag{8.2}$$

Based on the kinetic diameters of the gas molecules, CO_2 has a smaller kinetic diameter (3.30 Å) than CH_4 (3.80 Å); accordingly, CO_2 has a higher diffusivity than CH_4 [39]. Conversely, permeability depends upon gas solubility, i.e., the ability of the penetrant gases to condense. CO_2 has superior condensability than CH_4, as specified by the critical temperature "Tc," e.g., Tc, $CO_2 = 304$ K, Tc, $CH_4 = 191$ K, making it more permeable than CH_4. The transport properties of some gases are presented in the Table 8.1 [24]. Three main categories of non-porous membranes are polymeric membranes, inorganic membranes, and mixed matrix membranes, which are discussed below.

Membrane technology exploits the high solubility and fast permeability of tiny molecules like CO_2 and H_2S compared to the larger ones (N_2, heavier hydrocarbons including CH_4, etc.). When a CH_4/CO_2 gas mixture, like biogas, is fed into a membrane-based upgrade system, CO_2 will permeate faster into the membrane than other biogas constituents [7]. Thus, the high-pressure feed is separated into a CO_2-rich, low-pressure, permeate stream and a CH_4-enriched, retentate stream.

Table 8.1 Transport properties of gaseous constituents affecting solution-diffusion mechanism [6, 18, 32]

Gas	Kinetic diameter (nm)	Condensation temperature (K)
CO_2	0.330	304.21
CH_4	0.380	191.05
H_2	0.289	33.24
N_2	0.364	126.2
O_2	0.346	154.6

8.1.5 *Types of Non-porous Membranes*

8.1.6 *Polymeric Membranes*

The linkage of many small repeating units termed monomers forms bulky and complex molecules (macromolecules) known as polymers. Different states in a polymer are responsible for variations in its mechanical, thermal, and permeation properties. Before selecting a polymeric membrane, one must know about its Glass transition temperature and crystallinity. When an amorphous polymer is heated beyond its glass transition temperature, it transforms from a glassy state to a rubbery state. During this transition, polymer chains become flexible from a frozen state. In the state of flexibility or rubber form, selectivity is low, and permeability is high, whereas in frozen form or glass state, permeability is low, and the selectivity is high. Glassy polymers are frequently used in CO_2/CH_4 separation from a gas mixture because they have high permeability, exceptional thermal and chemical strength, are lightweight, and can be quickly processed into thin films, porous globules or beads, hollow fibers, and composites [13].

8.1.6.1 Types of Polymeric Membranes

Polymeric membranes dominate the membrane-based gas separation industry due to their low cost and excellent mechanical strength at high pressure. They can be quickly drawn into flat sheets and hollow fibers. These are highly selective and made very thin to achieve sensible capacities per unit area. Dense, non-porous membranes dominate the CH_4/CO_2 separation market [18]. The main polymers used for CO_2 and CH_4 removal are cellulose acetate (CA), polycarbonate (PC) polyesters (PE), polysulfone (PSF), polyimide (PI), polyetherimide (PEI), polypyrrole, cellulose acetate, polysulfone, and polyimide. CA and PI are the most preferred polymeric membranes for upgrading biogas [6]. CA is a rubbery polymer, the first commercial membrane to separate CO_2 and H_2S from gas mixtures. CA has the advantage of low cost due to the availability of cellulose in abundance but has the disadvantage of plasticization ($P_{\text{lasticisation}} = 8$ bar) due to the presence of its -OH-rich functional group, which quickly dissolves CO_2 within the membrane [39].

PI membranes are best for CO_2/CH_4 separation as compared to other PSP and PC. It has a crystal-like structure with high permeability and selectivity and excellent mechanical, thermal, and chemical stability. Despite its abundant availability, it is not cheap and vulnerable to plasticization ($P_{\text{lasticisation}} = 17$ bar) [6]. However, CA and PI membranes are sensitive to moisture. Hence, pre-removal of moisture is a prerequisite before upgrading biogas. The permeability and selectivity of polymeric membranes for CO_2/CH_4 separation are summarized in Table 8.1 (Table 8.2).

Table 8.2 Permeability and selectivity of polymer membranes for gas separation [6]

Polymer	Permeability at 30 °C (barrer; 1 barrer = 10^{-10} [cm^3 (STP)cm cm^{-2} s^{-1} $cmHg^{-1}$]	
	CH_4	CO_2
Cellulose acetate (CA)	0.21	6.3
Polyimide (PI)	0.25	10.7
Ethyl cellulose (EC)	19	26.5
Polycarbonate (PC)	0.13	4.23
Polydimethylsiloxane (PDMS)	800	2700
Polymethyl pentene (PMP)	14.9	84.6
Poly(p-phenylene oxide) (PPO)	11	75.8
Polysulfone (PSF)	0.25	5.6

8.1.6.2 Problems Associated with Polymeric Membranes

Polymeric membranes face two key issues: permeability/selectivity trade-off and plasticization at high pressure or long gas separation when CO_2 concentration is high. The permeability and selectivity have an inverse relation with each other, which has been reported by several studies [4, 19]. This trade-off is presented in Robeson's upper bound curve [7], which is a standard for gas membrane development.

Plasticization refers to the process in which the CO_2 permeability increases, and selectivity decreases as the pressure and concentration (partial pressure) increase beyond the critical level [6, 7]. The new generation of polymeric membranes maintains their strength in rigorous environments, i.e., they bear high mechanical stability at high inlet gas pressures/temperatures, have high resistance to plasticization, and have a persistently stable operation. Several methods, such as thermal treatment, cross-linking, and polymer blending, are available to suppress CO_2 plasticization. These methods are also effective in improving the performance of the membrane separation process.

8.1.7 Inorganic Membranes

Inorganic membranes have gained popularity in the last decade. Inorganic membranes offer more benefits regarding high mechanical strength and thermal and chemical stability than polymeric membranes. These perform well under high temperatures and possess more excellent chemical stability than polymeric membranes. Inorganic membranes provide permeability and selectivity beyond the Robeson upper bound. Under harsh conditions of high temperatures and pressures, inorganic membranes have five to ten times higher permeability-selectivity than polymeric membranes [7].

Inorganic membranes, whether porous or non-porous, are crucial in CO_2 separation. These membranes are made of various materials such as metal (alumina,

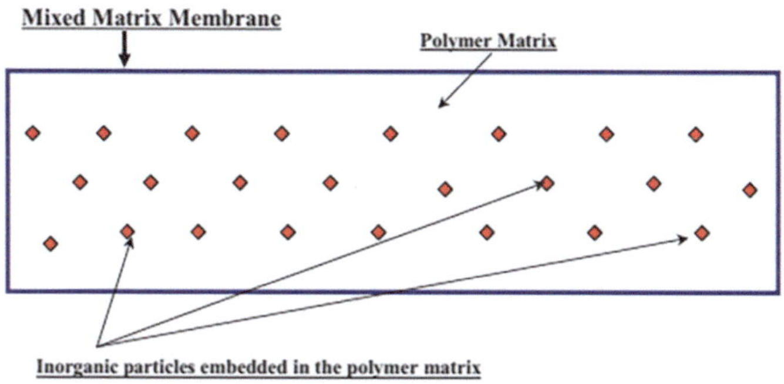

Fig. 8.2 Structure of mixed matrix membranes. *Reproduced with Permission from Elsevier* [3]

cobalt, copper, iron, nickel, niobium, palladium, platinum, tantalum, and vanadium), zeolites, carbon, and ceramic. Porous inorganic membranes, like zeolites, comprise polycrystalline films of alumina-silicates mounted on subnanometer porous supports. Inorganic membranes, including zeolites, activated carbon, silica, carbon nanotubes (CNT), and metal–organic framework (MOF), facilitate higher CO_2 separation and offer the advantage of chemical resistance to plasticization. Silica and carbonized membranes, for instance, separate CO_2 using the molecular sieving method [6].

8.1.8 Mixed Matrix Membranes

The limitations of polymeric and inorganic membranes led to the development of mixed matrix membranes (MMMs) [7]. High process performance, reasonable processing cost of base polymeric membranes, and better separation performance of inorganic particles are merged properties in MMMs. These integrate the advantages of polymers and inorganic membranes, such as low-cost and accessible synthesis of polymeric membranes combined with high selectivity and high permeability of inorganic membranes. MMMs are the solution for the trade-off problem experienced by polymeric membranes. MMMs are heterogeneous amalgamated hybrid membranes with continuous organic polymer matrix dispersed with micro or nano-scale inorganic filler particles, as shown in Fig. 8.2 [7].

8.1.9 Carrier (Facilitated Transport) Membranes

Recently, carrier membranes have gained importance, enabling high selectivity and permeability. Membrane separation also depends on the reactivity and affinity of the molecules toward the membrane. Firstly, the CO_2 reactive carriers react with the gas stream to produce carrier-CO_2 complexes. It is then followed by transferring the CO_2 carrier complexes through the membrane. Lastly, CO_2 is released from the complex on the low-pressure side. The selectivity of carrier membranes is determined not only by the material of the membrane itself but also by a carrier molecule that allows specific transport. This carrier can be combined with a membrane matrix in fixed form, and when it is dissolved in a liquid, it is in mobile form. A carrier-mediated CO_2-facilitated diffusion phenomenon is created along with a solution-diffusion mechanism by the carriers [39].

8.1.10 Membrane Modules

In recent years, membrane contactors have gained much attention due to effective gas separation because they combine gas absorption with membrane technology, and the concentration difference is the driving force. The non-porous hydrophobic membrane provides a large contact area per unit volume and performs gas/liquid mass transfer without dispersion of phases. A large surface area of the membranes is required for effective gas separation. An increase in surface area would increase the size of the membrane units. Membranes are, therefore, packed efficiently and economically into small sizes called membrane modules. Tubular, plate-and-frame, spiral wound, and hollow-fiber arrangement are available in different configurations. The membrane's permeability is inversely proportional to the membrane area required for gas separation [34].

8.1.11 Tubular and Capillary Modules

They are tube-like structures with porous walls. The shell comprises several membrane tubes assembled in a shell-and-tube type of arrangement. Thin membrane layers with a diameter typically greater than 10 mm are applied to the inside or outside of a tube support to form tubular membranes. The module may have four to eight tubes, but this is not the only possible number. The tube length ranges from 0.6 to 6 m, and the internal diameter varies from 5 to 25 mm [22]. The feed flows tangentially across the length of the membrane tube and gets filtered out into the outer shell. The retentate (concentrate) collects at the opposite end of the membrane tube, as shown in Fig. 8.3. These modules have the advantage of processing complex feed streams. Cleaning tubular modules is easy and can be done by mechanical or chemical methods. However, it has some drawbacks, such as low surface area to volume ratio, low packing density ($300m^2/m^3$), high holdup volumes, and ample space requirement. On the other hand, the capillary module is self-supporting, with the capillaries bound together at the free ends with agents such as epoxy resins, polyurethanes, or silicone rubber. These have high packing densities of 600–1200 m^2/m^3 [22]. Harasimowicz et al. [15] observed that using capillary modules with polyimide membranes

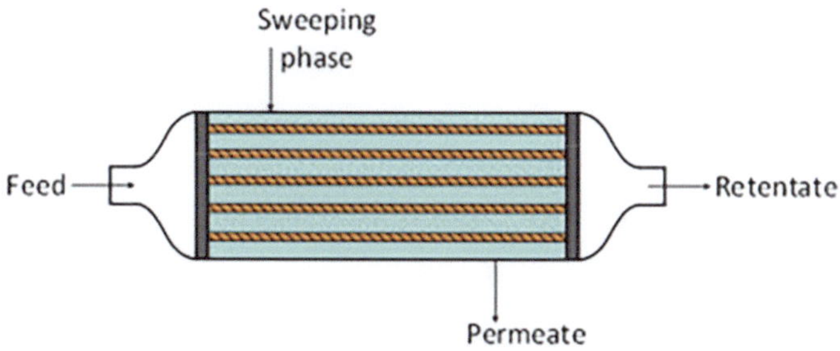

Fig. 8.3 Tubular module for membrane separation. *Reproduced with Permission from Elsevier* [29]

made it possible to achieve CH_4 concentrations up to 91–94.4% in the upgraded biogas.

8.1.12 Plate-and-Frame Modules

These are the oldest types of membrane systems developed, especially for small-scale applications. The membrane is sandwiched between two support plates, facilitating the fluid's flow parallel to both sides. The plate-and-frame module is illustrated in Fig. 8.4. These modules are expensive and have a lower packing density (100–400 m^2/m^3) than others [22]. The leaks through the gaskets are a severe problem in these modules. Spiral-wound modules and hollow-fiber modules essentially substitute these.

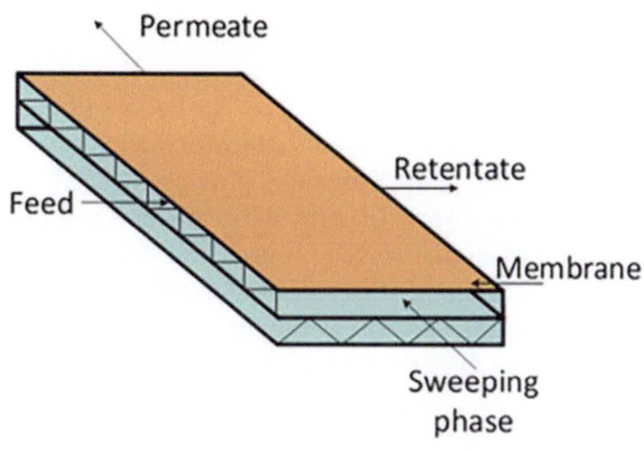

Fig. 8.4 Plate-and-frame module for membrane separation process. *Reproduced with Permission from Elsevier* [29]

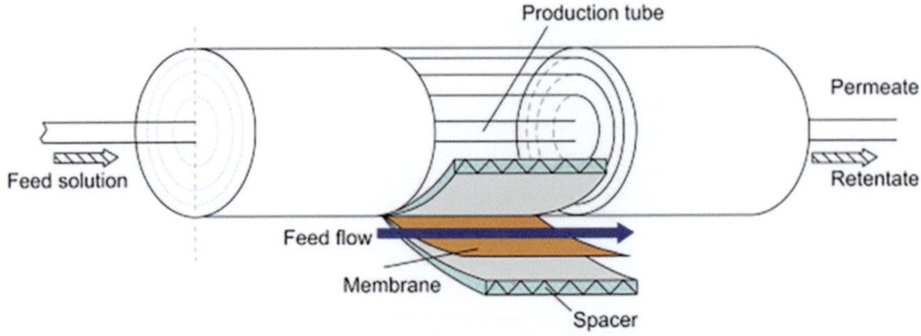

Fig. 8.5 Spiral module for membrane separation. *Reproduced with Permission from Elsevier* [29]

8.1.13 Spiral Modules

They contain several membrane envelopes draped parallelly around a central collection pipe. Multi-envelope designs are generally preferred to minimize the pressure drop of the permeate moving toward the pipe. Spiral modules, as shown in Fig. 8.5, offer high packing densities (300–1000 m^2/m^3) compared to envelope-type modules and are relatively cheaper [22].

8.1.14 Hollow-Fiber Modules

In a microporous hollow-fiber membrane contactor, CO_2 absorption is carried out by the countercurrent flow of the liquid absorbent and gas on the other side of the membrane wall. They offer the highest packing density (30,000 m^2/m^3) and contact surface with the lowest cost per unit area. Tiny tubes are packed closely together to maximize the contact surface area. The hollow-fiber configuration has a high specific area, and compactness is most considerable, as shown in Fig. 8.6 [11]. This configuration of membrane contactors provides 65–75% less weight and size than other conventional towers. Usually, these modules comprise 50–3000 individual hollow fibers, bundled and sealed on each end with epoxy in a hydraulically symmetric casing. The diameter and length of the fiber are in the range of 0.2–3 mm and 18–120 cm, respectively. These modules have good backwash capacity, which facilitates easy cleaning. Since these modules are susceptible to being blocked by large particles in the feed. Thus, pre-filtration to reduce the particle size to 100 μm is required. The liquid-Cel® Extra-Flow module is the most familiar design for hollow fibers and is based on a concentration gradient [12].

Cussler [9] reported 30 times higher efficiency of hollow-fiber contactor with major drawbacks of solvent-induced deterioration, membrane wetting (pores being filled with the adsorbent liquid, thereby increasing pore blocking and mass transfer

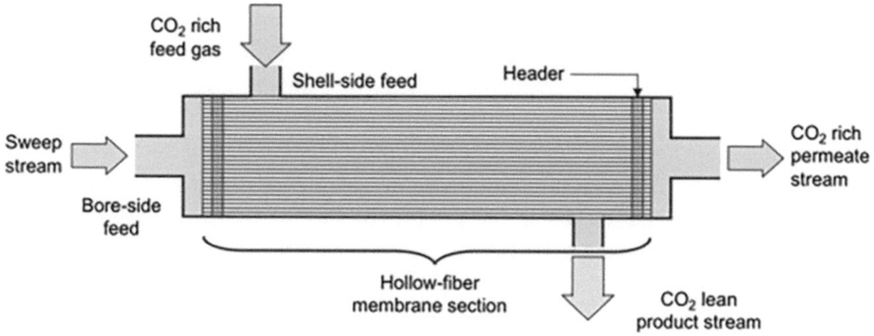

Fig. 8.6 Schematic diagram of hollow-fiber contactor. *Reproduced with Permission from Elsevier* [38]

resistance of the membranes). Chen et al. [7] and Park et al. [30] optimized hollow-fiber module configuration using polypropylene (PP) and water as absorbents for upgrading biogas. The configuration enhanced biomethane recovery by 90% and purity by 97%.

8.1.15 Membrane Separation Processes for Biogas Upgrading

One established and commercially accessible method is membrane separation, which can be used in low-pressure gas–liquid or high-pressure gas–gas modules. Nevertheless, its market share in the biogas upgrading sector is a mere 10% [36]. Biomethane in gas–liquid or multiple-stage gas–gas units with 96–98% CH_4 purity is guaranteed by membrane makers and process developers. To raise the methane concentration in the upgraded biogas, more energy must be used and sometimes a greater membrane area is needed. Varying designs and uses result in different methane recovery rates [7, 31].

The energy consumption of a membrane upgrading plant is mostly determined by the energy consumption of the compressor. Therefore, if represented as kWh/Nm^3 raw biogas, the energy consumption will be independent of the composition of the raw gas. The producers claim that they can guarantee electricity in the range of 0.20–0.30 kWh/Nm^3 [23]. Raw biogas capacity (Nm^3/h) can be assured. Regardless of size, this requirement holds true for the majority of applications. A given application's energy consumption will be determined by a number of factors, including the methane slip, the installed membrane area, the applied pressure, and the necessary carbon dioxide removal (i.e., the methane concentration of the biomethane produced). If a high methane concentration is needed in the biomethane, then a greater membrane area and higher pressure are necessary [26].

For membrane modules with capacities of 100–400 Nm^3/h, the investment costs range from 2500 € per Nm^3/h to 6000 € per Nm^3/h, accordingly. For plants with capacities greater than 1000 Nm^3/h, the capacity investment cost stays fixed at 2000 € per Nm^3/h [7].

Operating costs are mostly caused by the replacement of membranes, which typically last 5–10 years, the cost of pressurizing biogas, which uses energy between 0.2 and 0.38 kWh Nm^3, and the cost of pre-treating biogas. Membrane-based upgrading machines have maintenance expenses that are 3–4% greater than the initial investment [28]. A biogas upgrading unit's availability is often assured to be greater than 95%. Some membrane plants that are currently in operation have availability rates higher than 98%. Critical component redundancy can steadily boost availability while raising investment costs.

In addition to being chemically resistant to the significant levels of H_2S and H_2O contained in the biogas, the membranes used for the biogas upgrading process should be strong enough to withstand extreme process conditions like high pressure and temperature. To achieve low capital and operating expenses for biogas upgrading, membranes should have high CO_2 permeance and selectivity for CO_2/CH_4 in the range of 10–30. Better economics are offered by two-stage membrane designs [16].

Because of their comparatively low manufacturing costs and capacity to be processed into flat sheet and hollow-fiber configurations, polymeric membranes are the industry leader in biogas upgrading. Pre-conditioning the inlet gas stream with moisture is a prerequisite to sustaining the membrane's selectivity [37]. Table 8.3 compares various types of membranes used for CO_2 and CH_4 separation from the gaseous mixture.

There are two types of membrane separation processes. A low-pressure system with gas on one side and liquid on the other, and a high-pressure system with gas streams on both sides of the membrane wall. Below is a detailed explanation of these.

Table 8.3 Comparative summary between different types of membranes for biogas upgrading. *Reproduced with permission from the Publisher* [20]

Membrane classification	Membrane type	Energy requirement	Permeability-selectivity	Chemical requirement	Cost
Polymer	Gas permeation	High	High	Required	High
	Carrier	Low	Moderate	Not required	Low
	Hollow-fiber membrane contactor	Lowest	Highest	Required	Lowest
Inorganic	Inorganic	Highest	Highest	Not required	Highest
Mixed (hybrid)	Mixed matrix membrane	Moderate	Moderate	Not required	Moderate

8.1.16 Gas–Gas Absorption Method

A typical membrane-based biogas upgrade process is shown in Fig. 8.7. The following steps are generally employed for the upgrade process. The raw biogas is usually cleaned prior to the separation of CH_4 and CO_2 in a membrane system to remove water (to prevent clogging) by cooling and condensation and hydrogen sulfide (to prevent corrosion of compression equipment) with activated carbon. A particle filter is also used to protect the compressor and the membranes. Biogas is compressed to high pressures of approximately 20–40 bars depending upon the upgrading system's design and manufacturer and the upgraded gas's end use [34]. The membrane system is designed depending on the type of membranes being used, the quality of the product gas required, and the system's manufacturer. The gas–gas membrane system employs gas streams on both sides of the membrane wall. It is also called the high-pressure dry method and is one of the traditional approaches to CO_2 separation that uses membranes. When this system is used for biogas upgrading, pre-removal of moisture and H_2S from biogas is a requisite. This system operates at high pressures of 20–40 bar [7].

This method's limited CH_4 recovery is a significant drawback since CH_4 also flows through the pores under pressure. Biogas is upgraded to a maximum of 92% CH_4 in single-stage systems due to low CH_4 recovery [7]. Larger or multi-stage (two or three) membranes are used for upgrading to produce high-purity biomethane (over 95% CH_4). Thus, a balance between low CH_4 loss and high-purity biomethane must be struck.

Single or multiple membrane modules with internal permeant recirculation for CH_4 recovery are among the several configurations in which gas–gas units are created. Only the penetrated gas from the final module is recirculated when several units are connected in sequence. Multi-stage modules increase CH_4 recovery from 80% to 99.5%, according to [7]. In addition, compared to a single membrane, the multi-stage membrane method has the advantages of high CH_4 purity, biomethane recovery, and cheaper investment and operational costs [8].

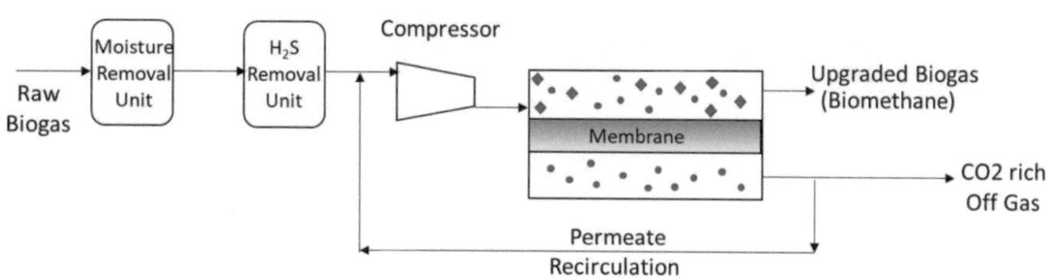

Fig. 8.7 Process flow diagram of the membrane separation process. *Reproduced with permission from the Publisher* [20]

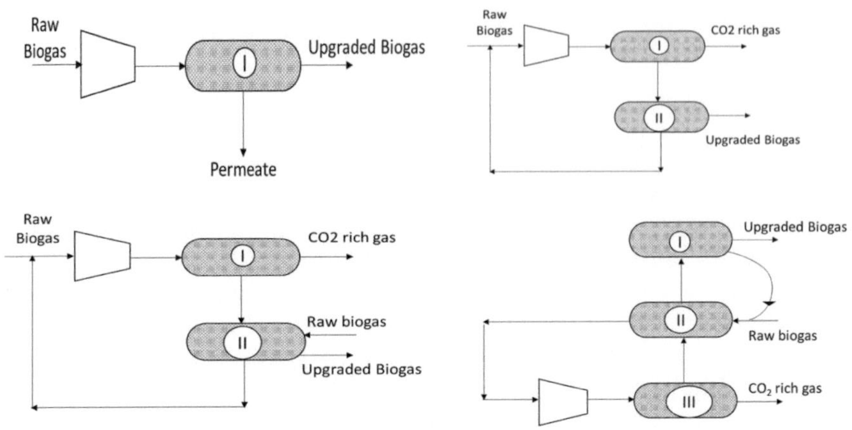

Fig. 8.8 **i** Single-stage configuration, **ii** two-stage configuration with a recirculation loop, **iii** two-stage configuration with sweep, and **iv** three-stage configuration with sweep. *Reproduced with permission from the Publisher* [20]

8.1.16.1 Single-Step Gas Permeation Processes

Two one-step membrane permeation procedures for biogas upgrading were presented by [34]. The single-stage gas process is the simplest gas permeation method for removing CO_2 from CH_4, as seen in Fig. 8.8i. Selectivity determines the high CH_4 losses that result from using this technique. Methane loss on the permeate side of these plants might range from 10 to 15%. The single membrane module procedure significantly increases CH_4 recovery on the right, which is combined with partial permeate stream recycling. Although a CH_4 purity of 96% is required for grid injection, CH_4 recovery cannot exceed 95%. Additionally, partial recycling raises the flow rate via the compressor, which raises the energy needed to power the separation process.

8.1.16.2 Multi-step Gas Permeation Processes

Using different membrane modules in a multi-stage system is a way to boost the upgrading system's CH_4 recovery while also achieving excellent product purity [21]. Figure 8.8ii and iii depicts two-stage upgrading processes. Two or three membrane separation units are used in the multi-stage processes to get high-quality upgraded biogas and high CH_4 recovery. The necessary CH_4 purity is achieved in a single stage, as illustrated in Fig. 8.8ii, which [10] examined. However, a second gas permeation module is used to recover the significant volumes of CH_4 that permeate through the membrane. A second compressor must produce the driving power of permeation for the second gas permeation module. In front of the first-stage compressor, the second module's CH_4-rich stream is recycled and combined with raw gas. Figure 8.8's process and the one in Fig. 8.8iii are connected ii. In order to achieve the necessary purity for grid injection, the CH_4-rich stream acquired from the second membrane stage is purified here. This stream is combined with the first stage's CH_4-rich stream.

As shown in Fig. 8.8iv, a sweep stream and compressing of raw gas are applied in a three-stage configuration to generate the driving force of permeation. Here, the raw biogas is used as a sweep gas to increase the driving force in the second membrane module [25]. Hence, a lower membrane area is needed without increasing the recycling stream [14]. Three-stage membrane-based biogas upgradation processes can obtain CH_4 purities of more than percent and CH_4 recoveries of more than 99% by optimizing operating pressure and flow rates in recycle streams, a balance achieved between the operating costs and the investment costs [33].

8.1.17 Gas–Liquid Absorption Method

Recently, a membrane-based gas–liquid absorption technique was developed for biogas upgrading. The benefits of membrane gas separation and chemical absorption process are integrated in gas-the gas–liquid (solvent) method. The microporous hydrophobic membrane with a large interfacial area separates the gas from the solvent flowing counter on each side of the membrane wall [11]. Physical and chemical solvents like water, seloxol, MEA, DEA, etc., can separate the membrane-based gas–liquid. Water as a liquid solvent in membrane separation gives somewhat lower purification rates. Raw biogas can be upgraded efficiently with more than 96% CH_4 in a single step with amine solutions. Pressurized gas molecules (CO_2 and H_2S) diffuse to the other side of the membrane wall and get absorbed by the solvent. Gas pressurization to low pressures (1–2 bar) prevents the solvent from flowing to the gas side. This process offers high gas separation efficiency of the gaseous constituents with low costs and energy requirements. Heating is used to regenerate amines, which often yields pure CO_2 and can be sold for use in industry [39]. Another important benefit of membrane technology is that when enough driving force is used for permeation, siloxanes, H_2S, and H_2O pass through the membrane more quickly than CO_2, making it easier to remove these components from biogas at the same time [2]. To prevent degradation, clogging, and to extend the membrane's life, biogas pre-treatment is strongly advised prior to membrane separation.

8.2 Conclusion

Membrane technology is a relatively recent technology but is very promising. Most membrane-based biogas upgrading plants use polymers. Further research should be focused on developing better membrane materials for higher performance with higher permeability, selectivity, and stability. The cost and lifetime of membranes are an issue that requires proper attention. Membrane-based biogas upgrading systems should be designed and staged to separate impurities like H_2S, siloxanes, and moisture simultaneously. For example, polydimethylsiloxane (PDMS) can separate siloxanes, while polyimides (PI) can separate CO_2. More research should be directed toward developing optimized membrane-based biogas upgrading systems, multi-stage membranes, and hybrid processes with low costs, easy operation, and enhanced energy efficiency.

References

1. Abetz V, Brinkmann T, Sozbilir M (2021) Fabrication and function of polymer membranes. Chem Teach Int 3(2):141–154. https://doi.org/10.1515/cti-2020-0023
2. Adewole JK, Ahmad AL, Ismail S, Leo CP (2013) Current challenges in membrane separation of CO_2 from natural gas: a review. Int J Greenh Gas Control 17:46–65. Elsevier Ltd. https://doi.org/10.1016/j.ijggc.2013.04.012
3. Aroon MA, Ismail AF, Matsuura T, Montazer-Rahmati MM (2010) Performance studies of mixed matrix membranes for gas separation: a review. Sep Purif Technol 75(3):229–242. Elsevier B.V. https://doi.org/10.1016/j.seppur.2010.08.023
4. Bahukudumbi P, Ford DM (2006) Molecular modeling study of the permeability–Selectivity trade-off in polymeric and microporous membranes. Ind Eng Chem Res 45(16):5640–5648. https://doi.org/10.1021/ie051366t
5. Baker RW (2012) Membrane technology and applications. John Wiley & Sons
6. Basu S, Khan AL, Cano-Odena A, Liu C, Vankelecom IFJ (2010) Membrane-based technologies for biogas separations. Chem Soc Rev 39(2):750–768. https://doi.org/10.1039/B817050A
7. Chen Q, Acey C, Lara JJ (2015) Sustainable Futures for Linden Village: a model for increasing social capital and the quality of life in an urban neighborhood. Sustain Cities Soc 14(1):359–373. https://doi.org/10.1016/J.SCS.2014.03.008
8. Chmielewski AG, Urbaniak A, Wawryniuk K (2013) Membrane enrichment of biogas from two-stage pilot plant using agricultural waste as a substrate. Biomass Bioenerg 58:219–228. https://doi.org/10.1016/j.biombioe.2013.08.010
9. Cussler EL (2009) Diffusion mass transfer in fluid systems. www.cambridge.org/cussler
10. Deng L, Hägg MB (2010) Techno-economic evaluation of biogas upgrading process using CO_2 facilitated transport membrane. Int J Greenhouse Gas Control 4(4):638–646. https://doi.org/10.1016/J.IJGGC.2009.12.013
11. Dindore VY, Brilman DWF, Feron PHM, Versteeg GF (2004) CO_2 absorption at elevated pressures using a hollow fiber membrane contactor. J Membr Sci 235(1–2):99–109. https://doi.org/10.1016/j.memsci.2003.12.029
12. Gabelman A, Hwang ST (1999) Hollow fiber membrane contactors. J Membr Sci 159(1–2):61–106. https://doi.org/10.1016/S0376-7388(99)00040-X
13. Gholami M, Mohammadi T, Mosleh S, Hemmati M (2017) CO_2/CH_4 separation using mixed matrix membrane-based polyurethane incorporated with ZIF-8 nanoparticles. Chem Pap 71(10):1839–1853. https://doi.org/10.1007/s11696-017-0177-9
14. Haider S, Lindbråthen A, Hägg MB (2016) Techno-economical evaluation of membrane based biogas upgrading system: a comparison between polymeric membrane and carbon membrane technology. Green Energy Environ 1(3):222–234. https://doi.org/10.1016/J.GEE.2016.10.003
15. Harasimowicz M, Orluk P, Zakrzewska-Trznadel G, Chmielewski AG (2007) Application of polyimide membranes for biogas purification and enrichment. J Hazard Mater 144(3):698–702. https://doi.org/10.1016/j.jhazmat.2007.01.098
16. Havas D, Lin H (2017) Optimal membranes for biogas upgrade by removing CO_2: high permeance or high selectivity? Sep Sci Technol (Phila) 52(2):186–196. https://doi.org/10.1080/01496395.2016.1218515

17. Ismail AF, Khulbe KC, Matsuura T (2015) Gas separation membranes polymeric and inorganic
18. Jeon YW, Lee DH (2015) Gas membranes for CO_2/CH_4 (biogas) separation: a review. Environ Eng Sci 32(2):71–85. Mary Ann Liebert Inc. https://doi.org/10.1089/ees.2014.0413
19. Kadirkhan F, Goh PS, Ismail AF, Wan Mustapa WNF, Halim MHM, Soh WK, Yeo SY (2022) Recent advances of polymeric membranes in tackling plasticization and aging for practical industrial CO_2/CH_4 applications—A review. Membranes 12(1). MDPI. https://doi.org/10.3390/membranes12010071
20. Kapoor R, Ghosh P, Kumar M, Vijay VK (2019) Evaluation of biogas upgrading technologies and future perspectives: a review. Environ Sci Pollut Res 26(12):11631–11661. Springer. https://doi.org/10.1007/s11356-019-04767-1
21. Kim KH, Baik KJ, Kim IW, Lee HK (2012) Optimization of membrane process for methane recovery from biogas. Sep Sci Technol (Phila) 47(7):963–971. https://doi.org/10.1080/01496395.2011.644878
22. Kluiters SCA (2004) Status review on membrane systems for hydrogen separation
23. Kvist T (2012) Membranes for upgrading biogas to natural gas quality. www.dgc.dk
24. Lin H, Freeman BD (2005) Materials selection guidelines for membranes that remove CO_2 from gas mixtures. J Mol Struct 739(1–3):57–74. https://doi.org/10.1016/j.molstruc.2004.07.045
25. Makaruk A, Miltner M, Harasek M (2010) Membrane biogas upgrading processes for the production of natural gas substitute. Sep Purif Technol 74(1):83–92. https://doi.org/10.1016/J.SEPPUR.2010.05.010
26. Mamone RM (2014) Small-scale biogas upgrading with membranes: a farm based techno-economic and social assessment for sustainable development
27. Masebinu SO, Aboyade AO, Muzenda E (2014) Operational study and simulation of a biogas upgrading plant. In: Newswood Limited : international association of engineers
28. Muñoz R, Meier L, Diaz I, Jeison D (2015) A review on the state-of-the-art of physical/chemical and biological technologies for biogas upgrading. Rev Environ Sci Biotechnol 14(4):727–759. Springer Netherlands. https://doi.org/10.1007/s11157-015-9379-1
29. Nagy E (2019) Membrane materials, characterization, and transport properties. Basic EquS Mass Transp Membr Layer, pp 1–10. https://doi.org/10.1016/B978-0-12-813722-2.00001-7
30. Park A, Kim YM, Kim JF, Lee PS, Cho YH, Park HS, Nam SE, Park YI (2017) Biogas upgrading using membrane contactor process: Pressure-cascaded stripping configuration. Sep Purif Technol 183:358–365. https://doi.org/10.1016/J.SEPPUR.2017.03.006
31. Patterson T, Esteves S, Dinsdale R, Guwy A (2011) An evaluation of the policy and techno-economic factors affecting the potential for biogas upgrading for transport fuel use in the UK. Energy Policy 39(3):1806–1816. https://doi.org/10.1016/j.enpol.2011.01.017
32. Rufford TE, Smart S, Watson GCY, Graham BF, Boxall J, Diniz da Costa JC, May EF (2012) The removal of CO_2 and N_2 from natural gas: a review of conventional and emerging process technologies. J Petrol Sci Eng 94–95:123–154. https://doi.org/10.1016/j.petrol.2012.06.016
33. Scholz M, Alders M, Lohaus T, Wessling M (2015) Structural optimization of membrane-based biogas upgrading processes. J Membr Sci 474:1–10. https://doi.org/10.1016/j.memsci.2014.08.032
34. Scholz M, Melin T, Wessling M (2013) Transforming biogas into biomethane using membrane technology. Renew Sustain Energy Rev 17:199–212. https://doi.org/10.1016/j.rser.2012.08.009
35. Shindo R, Nagai K (2013) Gas separation membranes. In: Kobayashi S, Müllen K (eds) Encyclopedia of polymeric nanomaterials. Springer, Berlin, Heidelberg. https://doi.org/10.1007/978-3-642-36199-9_134-1
36. Thrän D, Krautz A, Scheftelowitz M, Lenz V, Liebetrau J, Daniel-Gromke J, Zeymer M, Nelles M (2014) Auswirkungen der gegenwärtig diskutierten Novellierungs-vorschläge für das EEG-2014. www.dbfz.de
37. Vrbová V, Ciahotný K (2017) Upgrading biogas to biomethane using membrane separation. Energy Fuels 31(9):9393–9401. https://doi.org/10.1021/acs.energyfuels.7b00120
38. Wang Z, Dong S, Li N, Cao X, Sheng M, Xu R, Wang B, Wu H, Ma C, Yuan Y, Xu H, Chen F, Zhao S (2018) CO_2-selective membranes: how easy is their moving from laboratory to industrial scale? In: Current trends and future developments on (Bio-) membranes: carbon

dioxide separation/capture by using membranes, pp 75–102. https://doi.org/10.1016/B978-0-12-813645-4.00003-9

39. Zhang Y, Sunarso J, Liu S, Wang R (2013) Current status and development of membranes for CO_2/CH_4 separation: a review. Int J Greenh Gas Control 12:84–107. Elsevier Ltd. https://doi.org/10.1016/j.ijggc.2012.10.009

Chapter 9
Developing Technologies for Biogas Upgrading

9.1 Introduction

Biogas upgrading is an area with much scope for new developments and improvements. To increase productivity and cost-effectiveness, there are chances to create new technologies and investigate various hybrid approaches—the fusion of established and cutting-edge technology. Recent developments in biogas upgrading technology include cryogenic, biological, and hybrid technologies. Most currently available processes are under development, so minimum data is available. Although promising, these techniques are still in the early stages of research and show superior performance at laboratory scales. For biogas upgrading technology to be used more widely worldwide, both established and new technologies are constantly being upgraded for higher performance, increased efficiency, and low cost.

9.2 Cryogenic Technology for Biogas Upgradation

Cryogenic technology has been well-established for natural gas purification for many years, but biogas upgrading and liquefaction are still nascent [31]. Many recent advances have been made in developing cryogenic technology for biogas upgrading since it is not a regular gas upgrading technique. For biogas, cryogenic technology is a one-stop solution for:

- Biogas Cleaning: removal of trace contaminants like H_2 S, etc.
- Biogas upgrading removal of CO_2, $N_{2,}$ etc.
- Liquefaction: condensation of upgraded biomethane to liquefied biomethane.

V. K. Vijay et al., *Compressed Biogas (CBG)*, Green Energy and Technology,
https://doi.org/10.1007/978-981-96-5697-4_9

Table 9.1 Condensation temperature for different gas constituents in biogas [15]

Compound	Temperature of condensation (1 bar) (°C)
Methane (CH_4)	−161.5
Carbon dioxide (CO_2)	−78 (sublimates)
Nitrogen (N_2)	−196
Oxygen (O_2)	−183
Hydrogen sulfide (H_2 S)	−60

9.2.1 *Technical Background*

The cryogenic process involves fractional condensation and distillation of gases at low temperatures. The process operates at low temperatures of −50 °C when common gases transform into liquid [16]. The different condensation temperatures of the gaseous components in biogas are presented in Table 9.1. CO_2 condenses at lower pressure and higher temperatures than CH_4, facilitating both separations.

In the cryogenic process, low temperature is required for upgradation. Two cooling options are available for achieving low temperatures: indirect cooling and direct cooling. A heat exchanger with liquid nitrogen accomplishes indirect cooling as a cooling medium to liquefy the biogas. However, due to the high running cost of nitrogen, it is not a feasible option for large-scale applications. Direct cooling involves cooling equipment like compressors, heat exchangers, and expansion to reduce temperature. Direct and indirect cooling is mainly used for biogas upgrading [19].

The working principle for liquefaction of biogas can be either closed-loop or opened-loop. In a closed-loop cycle, biogas is initially cooled by a separate refrigeration process using nitrogen as a cooling medium and subsequently expanded through either a valve or a turbine. This causes a reduction in pressure and temperature and, subsequently, in condensation of CH_4. High-pressure compression of biogas is followed by heat exchange with lower-temperature heat sinks, such as outside air or an external refrigeration cycle, in the open-loop cycle. Cooled biogas is then expanded through a turbine where the pressure and the temperature are low enough to condense or de-sublimate CO_2 [17].

9.2.2 *Biogas Upgrading Based on Cryogenic Technology*

The idea behind cryogenic biogas separation is that CH_4, CO_2, H_2 S, and N_2 condense or liquefy at various temperature–pressure thresholds. This facilitates the separation of CH_4 from other biogas components. Biogas is compressed and cooled by heat exchangers until liquefied CO_2 is separated from the remaining gas, producing highly pure biomethane. This is accomplished by successive temperature reductions to remove each contaminant in different steps. First, the temperature is reduced to

−25 °C, where H_2 O, H_2 S, and siloxanes are condensed. The temperature is then lowered to 55 °C, and the pressure is raised to 40 bars. CO_2 gets liquefied in this stage. To extract the remaining CO_2 as a solid and create pure biomethane, the temperature is then lowered to −85 °C. After that, the high-pressure CH_4 product gas can be depressurized for injection into a pipeline. As an alternative, the biomethane that is produced can be liquefied by further cooling it down [11].

The liquefied CO_2 produced in the second phase can be marketed as high-purity byproducts to improve the overall economic process performance. Because hydrogen sulfide has a greater temperature than CO_2, it can also be eliminated. Since nitrogen and oxygen have lower boiling temperatures than CH_4 (−196 °C and −183 °C, respectively, at 1 bar), they are not extracted directly from the biogas. Water vapor and H_2 S are eliminated before the cryogenic procedure to prevent freezing and corrosion issues [29]. High pressure prevents CO_2 from freezing, clogging the pipe and heat exchanger system, and keeping it in the liquid phase. Several compressors and heat exchangers are used to maintain these operating requirements.

The liquid state of CO_2 is impossible at pressures below 5.1 atm (~5.2 bar). Instead of melting, solid CO_2 sublimates straight into its gas phase when heated to −78.5 °C at atmospheric pressure [18]. Due to this characteristic, the solid phase is typically called dry ice, it does not melt.

9.2.2.1 Cryogenic Distillation

The most popular method for improving biogas is cryogenic distillation. This technique can be carried out using flash, single-column, double-column, two-column, and three-column systems, among other approaches. Figure 9.1 illustrates cryogenic distillation, which condenses and freezes CO_2 before CH_4. In this case, the CO_2 can be extracted either liquid or solid. Biogas is cooled during the procedure before it enters into the distillation column. For the various gaseous components to liquefy, the gas is typically first compressed between 18 and 30 bar and then cooled via a heat exchanger to −45 to −59 °C. Increased pressure prevents carbon dioxide from sublimating. A condenser extracts the product from the top of the distillation column, which includes most of the CH_4 [6, 7].

[42] studied a biomethane purity of 94.5 mol.% CH_4 and liquid CO_2 with a purity of 99.7% can obtained by combining distillation, flash separation, auxiliary refrigeration, and internal heat recovery. The energy requirement of the process with such a combination was 0.26 kW h/Nm^3 raw biogas. Yousef et al. [41] varied temperature, distillation pressure, biogas composition, reflux ratio, number of trays, and CH_4 concentration in biomethane to control frosting and decrease the energy requirement of the process. CH_4 purity of 97.12%, along with a liquid CO_2 having 99.92% purity, was achievable using two distillation columns.

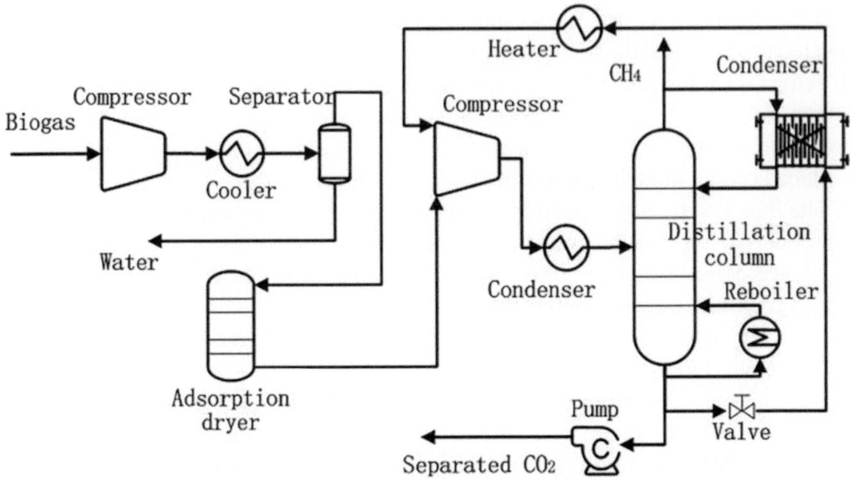

Fig. 9.1 Process flow diagram for cryogenic distillation. *Reproduced with Permission from Elsevier* [38]

9.2.2.2 Cryo Pur System

This process was developed by Cryo Pur company based in France. The Cryo Pur system is an integrated cryogenic process and a liquefaction step. This process is suitable for large-scale systems, with gas streams having high concentrations of N_2 and O_2, like landfill biogas. According to the company, the benefits include low power consumption, high heat recovery, a liquid carbon dioxide purity above 99.99%, and a liquid biomethane purity above 99%. Initially, biogas is pre-treated using activated carbon to remove impurities like H_2 S, siloxanes, etc. It is followed by gas cooling to −90 °C to frost and defrost water in a heat exchanger [35].

After that, CO_2 is captured and separated in liquid form at −120 °C by a two-heat exchanger subsystem, allowing it to be employed immediately as a liquefied product. Bio-LNG can be kept once biomethane is compressed and cooled at 15 bar and −120 °C, respectively. A 96–97% methane product with 0.5–3% methane slip can be produced by cryogenic distillation. The main advantage of cryogenic separation is the large quantities and the high purity of the liquid CO_2 and biomethane. The yield of this separation is 99% for CH_4.

9.2.2.3 Advantages and Disadvantages of Cryogenic Techniques

This process is ideal for large-scale systems with high concentrations of CO_2, as found in wetlands, landfills, sewers, etc. Because nitrogen gas can be isolated during the CH_4 liquefaction process, this technique is appropriate for upgrading landfill gas. Cryogenic technology can be used where no gas grid is present, reducing limitations due to location. A significant amount of extra heat is generated during the energy-intensive process of liquefying natural gas; this heat can be recovered as thermal or electric energy and used elsewhere. Cryogenic technology, in general, is costly but very energy efficient. It does not require water or any chemical to remove impurities. This method produces upgraded biogas that is already at a high pressure and doesn't require additional compression to be used as fuel for vehicles. Liquid CO_2 is obtained as a by-product that could be sold in solid form to increase the economics of the process.

The main disadvantage of cryogenic separation is that the cryogenics process uses a complex process layout. Compressors, turbines, and heat exchangers increase complexity and capital costs. The final price of upgraded biogas using this technique is very high. The process requires high pressure, low temperature, and adequate insulation, which poses many operational difficulties. The electricity demand for the process ranges from 0.63–1.8 kWh of electricity per Nm^3 of biogas for upgrading, which is very high. Sometimes, frost is produced by CO_2, which causes difficulty in operating the heat exchanger. Cryogenic distillation systems are expensive, even though they require less maintenance [35].

9.2.2.4 Case Study

According to numerous research studies on upgrading biogas, the cryogenic process is still relatively new, but some businesses have already used it in pilot plant phases. Among the businesses that use this technology are Gas Treatment Services B.V. from the Netherlands, Scandinavian Gts, Acrion Technologies, and Prometheus-Energy. There are a few biogases and cytotechnology pilot plants in the world. In 2004, Acrion constructed a pilot operating for six months at Burlington County Landfill in New Jersey, USA. The purpose of the pilot was to evaluate their technique for producing LBG and purifying landfill gas using their CO_2 scrubber. The trials successfully produced 1.32 m^3 LBG/day while the pilot operated at 100 Nm^2 raw gas/h [17]. Prometheus constructed a pilot plant in 2000 that purified and liquefied landfill gas in a British Columbia, Canada landfill. Although the pilot was conducted in 2001, no data regarding the trials' outcomes was collected.

9.3 Biological Biogas Upgrading Technologies

Biological biogas upgrading uses microbes to convert CO_2 and H_2 into methane, as opposed to energy-intensive biogas upgrading technologies such as membrane separation, PSA, and water scrubbing [19]. Biological biogas upgrading involves two metabolic processes (Fig. 9.2). One is the direct conversion of CO_2 to CH_4 by hydrogenotrophic methanogens. The Sabatier process is a catalytic chemical reaction that can perform the precise conversion. The need for mild temperatures and greater resilience to gas contaminations such as H_2S, organic acids, and NH4 are only two

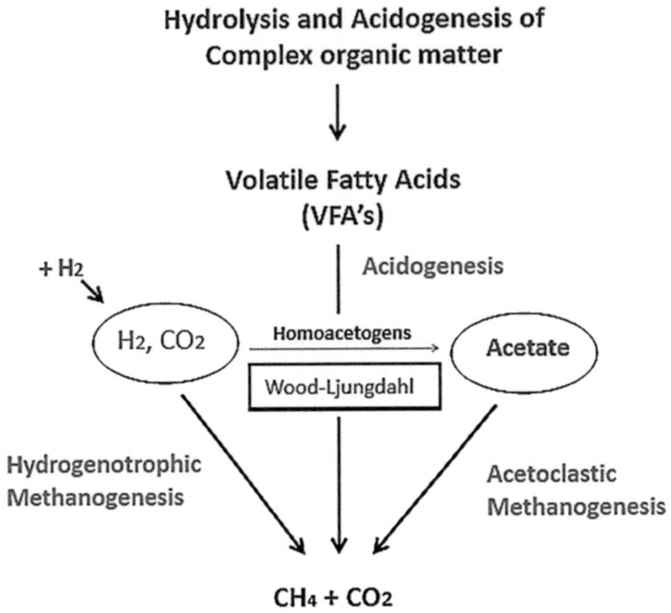

Fig. 9.2 Bacteria-mediated metabolic pathways for hydrogen-assisted methanogenesis. *Reproduced with permission from the Publisher* [19]

benefits biological methanation offers over chemical procedures [19]. Through the Wood-Ljungdahl pathway, the homoacetogenic bacteria first convert CO_2 to acetate, which is subsequently transformed into methane by the acetoclastic methanogens. This second metabolic pathway is an indirect biogas upgrade [4].

The hydrogenotrophic methanogen population is essential for efficient biogas upgrading, and there are several ways to boost their abundance, including endogenous enrichment with hydrogenotrophic methanogens or bioaugmentation with pure microbial cultures. Methanotrophic methanogens from the genera Methanobacterium, Methanoculleus, Methanomicrobium, and Methanothermobacter are more prevalent than acetoclastic methanogens like Methanosarcina, according to an assessment of the microbial community during biogas upgrading [1, 27].

9.3.1 Methods for Biological Biogas Upgradation

Biological biogas upgrading techniques come in three distinct types: (a) in situ technique, which introduces H_2 into a biogas reactor's liquid stage and causes intrinsic CO_2 to couple with the supplied H_2 producing methane; (b) ex situ technique, which introduces CO_2 and H_2 from outside sources into a reactor's liquid stage that contains hydrogenotrophic microorganisms, converting to CH_4 [19].

9.3.1.1 In Situ Upgradation Technique

While utilizing the current digester minimizes costs, numerous technological challenges accompany with in situ technique. These factors comprise: (a) a pH elevation exceeding 8.5 due to the depletion of bicarbonate, a crucial buffering agent, which impedes methane synthesis; (b) hydrogen's gas–liquid mass exchange; and (c) excessive hydrogen infusion in the reactor obstructs anaerobic breakdown and decreases pH due to the accumulation of electron accepting compounds, including volatile fatty acids (VFAs) and ethanol [23, 26, 39]. In addition, as H_2 regulates methanogenesis and may impact upstream processes, process optimization becomes more complicated. Under typical circumstances, the oxidation of volatile fatty acids (VFAs) is endergonic and only feasible when hydrogenotrophic methanogens are used to maintain a low partial pressure of hydrogen [14, 32]. However, upon the addition of $H_{2,}$ its partial pressure rapidly grows, inhibiting VFA oxidation and possibly causing the process to fail. Acetate can also be formed by activating homoacetogens; the process will stop if the rates of acetate generation and consumption are not equal. These constraints restrict in situ biogas upgrading to a limited number of research conducted on a lab scale [24, 26, 40]. More investigation and development are needed to understand the effects of an ideal and surplus supply of H_2 on the intricate biochemical processes of methanogenesis, homoacetogenesis, and VFA oxidation.

9.3.1.2 Ex Situ Upgradation Technique

The ex situ technique was developed to get around these obstacles and prevent core biogas production from being impeded. Here, biogas, syngas, and flue gas can be used as sources of CO_2, and hydrolysis of water utilizing renewable electricity sources, such as solar cells and windmills, can yield H_2. The ex situ method employs externally furnished CO_2 and H_2 in a separate reactor that contains enhanced blended or unadulterated hydrogenotrophic methanogenic organisms that use carbon dioxide and hydrogen to generate CH_4 [5]. According to [4], this method's primary benefits include the ability to maintain an extensive amount of influent gas for a shorter period of time, the ability to make use of unused carbon dioxide from additional sources, like syngas, etc., and the process's independence from biomass and simplicity in terms of biochemistry, which ensures the stability of the current biogas plant because it occurs in a separate reactor. Here, a variety of reactor types, including Bubble column, Series Flow Reactors, "*Continuous Stirred Tank Reactors* (CSTR)," *Fixed-Bed Reactors*, "*Anaerobic Trickle-Bed Reactors*" (ATBR), can be applied to overcome the restriction on H2 mass transfer from gas to liquid [20, 21]. Due to enhanced gaseous recirculation, [20] obtained 98% enriched methane in bubble columns and series flow reactors when investigating ex situ biogas upgrading utilizing three distinct bioreactor configurations. Additionally, the presence of novel uncultured phylotypes during H_2 -aided methanogenesis was discovered by a 16S rRNA-based microbial community study of the anaerobic digestion process. The predominant archaeal community on the bubble column reactor's surface was determined to be Methanothermobacter thermautotrophicus. The methane evolution rate (MER) in ex situ biological hydrogen methanation is higher than in situ biological hydrogen methanation, ranging from 0.37 to 688.6 L/Lreac/d [25, 26, 28, 40]. Making use of a methanogenic mixed culture at 800 rpm and operating in thermophilic conditions, CSTR produced a 95.4% methane enrichment [25]. ATBR demonstrated 98% CH_4 enrichment by liquid recirculation under mesophilic conditions while operating continuously with bound methanogenic microorganisms. Ninety-eight % of CH_4 concentrations were reported by ATBR when it was operating in thermophilic circumstances. In a separate study, 96% CH_4 synthesis was achieved by directly pumping biogas and hydrogen into the ATBR [30]. In Avedøre, Denmark, a commercial ex situ biological CH_4 upgrading factory called Electrochaea was founded [8]. Excess wind energy was used to power the electrolysis of water. A membrane-cleaning unit enhances the plant's 90–95% CH_4 concentration [5]. Recent studies on in situ and ex situ techniques are compiled in Table 9.2.

9.4 Hybrid Technologies

A hybrid system is created by combining the advantages and disadvantages of different biogas upgrading systems [31]. For instance, combining membrane gas permeation with cryogenic separation, amine absorption, or water scrubbing to create a hybrid process combines the benefits of each method. It enhances biogas upgrades' effectiveness and performance. Seven hybrid membrane process setups for biogas upgrading were examined by [33]. Membranes in conjunction with water scrubbing and cryogenic techniques have been shown to lower the energy intensity and upgrading costs of traditional separate technologies. Pentair has brought hybrid technology—which combines membrane and cryogenic technology—to the market. Temperature-membrane-cryogenic methods used less energy than separate technologies, according to [36]. In order to process biogas from the wastewater treatment plant, [34] combined temperature-swing adsorption with a membrane technique. By recycling the off-gas into the membrane process, it decreased CH_4 loss and produced biomethane with a 97% CH_4 purity. Hybrid biological upgrading, which combines in situ and ex situ biogas upgrading technologies, was recently studied by [13]. In the first phase, an in situ process upgrades the biogas to 80–90% CH_4 and captures some of the CO_2. This is followed by an ex situ process that further upgrades the enriched biogas to a CH_4 content greater than 95%. Enhancing overall system stability was a major difficulty in scaling up the hybrid upgrade system. To improve the techno-economic aspect of biogas upgrading, more hybrid technologies must be investigated in order to combine the benefits of two or more technologies.

Table 9.2 Comparative analysis of in situ and ex situ techniques for biogas upgradation. *Reproduced with permission from the Publisher* [19]

Category	Class of reactor	Feed material	Temperature (°C)	Retention time	Methane (%)	Assessments	References
In situ	Batch	Maize	52	–	89.4	The results showed that too much H_2 addition hinders methanogenesis and encourages homoacetogens. Controlling the H_2 input rate and gas retention period in relation to the CO_2 generation rate, hydrogen uptake rate, and development of hydrogenotrophic methanogens is essential to achieving a higher CH_4 content without the buildup of acetate and other VFAs	[27]
In situ	Batch	Sludge from wastewater treatment plant	35 and 55	15 d	96	For 96% H_2 conversion to CH_4 in 15 days, gas recycling at 100 L/L/d under thermophilic conditions was the optimum choice	[43]
In situ	CSTR	Biogas sludge	38	20 d	100	The rate of H_2 uptake and certain hydrogenotrophic methanogens were both successfully raised by pulse H_2 injections	[1]
In situ	Continuous	Chicken manure	–	10 d	60–70	Greater methane enrichment resulted from a greater effluent recirculation rate (200%–400%)	[12]
In situ	UASB	Potato starch wastewater	55	5–20 d	82	Porous inert devices, such as ceramic sponges, enhance H2 diffusion. Additionally, it was discovered that chamber configuration and gas recirculation maximized the synthesis of CH_4 from CO_2	[9]
Ex situ	Membrane bioreactor	Anaerobic sludge	55 ± 1	–	–	Gas sparging through the ceramic MBR demonstrated a significant H2 mass transfer capacity	[2]

(continued)

Table 9.2 (continued)

Category	Class of reactor	Feed material	Temperature (°C)	Retention time	Methane (%)	Assessments	References
Ex situ	CSTR, two reactors with upflow, and a batch and continuous bubble column reactor	Slurry treated via a biogas plant	52 ± 1	One by four to three by four days	79–98	Ex situ biogas upgrading and enhancing tests were conducted using various reactor setups. The output gas of flow reactors in series or bubble column reactors has the highest methane percentage (>98%). Additionally, it was noted that the efficiency of biogas upgrading was enhanced by raising the gas recirculation rate	[20]
Ex situ	UASB Continuous	Slurry treated via a the biogas plant	55 ± 1	Between 4 and 15 h	89.5–96.3	Gas–liquid contact was enhanced by the diffuser's larger pore size and increased gas recirculation rate	[10]
Ex situ	Trickle bed	Anaerobic culture that is blended	37 ± 2	Three and a half hours	96	An immobilized hydrogenotrophic microbial culture in a mesophilic trickle-bed reactor produced a CO_2 conversion of > 96%, CH_4 concentrations above 96%, and less than 0.1% residual H_2. There is no need for additional CO_2 removal because the gas quality meets grid requirements	[30]

9.4.1 Membrane-Based Hybrid Processes

According to [19], hybrid processes that combine membrane technology with traditional gas separation equipment may be better than individual procedures that use just one technology, such as membrane technology or amine scrubbing. Only moderate CH_4 purities are obtained from the membrane module in procedure (a) because the gas permeation module separates CO_2 and CH_4 in bulk. The CH_4 stream is polished by the traditional process apparatus downstream of the gas permeation module. There are two CO_2-enriched exhaust gas streams produced. The gas permeation module is used in procedure (b) to achieve CH_4 purity in a single membrane step. Nonetheless, substantial amounts of CH_4 are present in the permeate stream and ought to be collected. The traditional gas separation apparatus receives the permeate stream from the membrane stage. Since the permeate stream typically makes up less than 50% of the raw gas stream, the equipment size is minimal in this case [34]. The product from the gas permeation module can be combined with the recovered CH_4 under pressure. In order to produce pipeline-quality methane, [34] developed a two-stage membrane process in conjunction with temperature-swing adsorption (TSA) as a pre-treatment. Recycling the off-gas streams improves energy efficiency and CH_4 recovery.

9.4.2 Cryogenic-Based Hybrid Processes

(a) *Cryogenic Adsorption Process*

The cryogenic adsorption-based process, as shown in Fig. 9.3, integrates the adsorption-based biogas upgrading process with cryogenic technology. It gives an 88.9% recovery of CH_4 along with a high-purity liquid stream of CO_2 [22].

(b) *Cryogenic Membrane Process*

As shown in Fig. 9.4, this process integrates the advantages of cryogenic and membrane technologies in one hybrid system. These hybrid processes report reduced cost per CO_2 ton with 85% CO_2 capture and low investment costs [3, 33].

(c) *Cryogenic Hydrate Process*

Cryogenic hydrate processes, as shown in Fig. 9.5, are another potential technology that can be explored for biogas upgrading. Integrating these two technologies

Fig. 9.3 Flow process of the cryogenic adsorption-based process. Adapted from [6, 7]

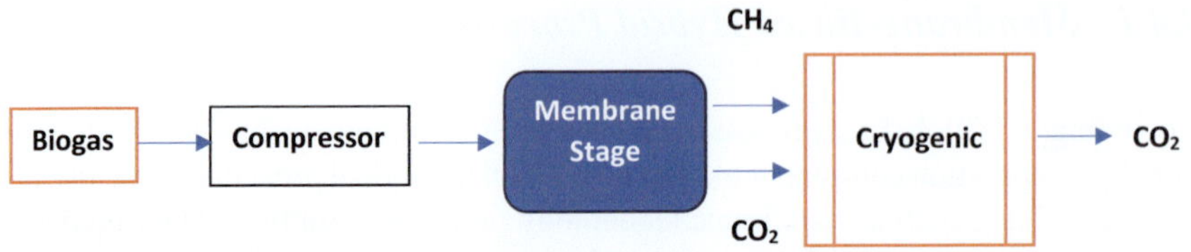

Fig. 9.4 Flow process of the cryogenic membrane process. Adapted from [6, 7]

provides optimum conditions for operating at low temperatures and high pressure. The cryogenic stage operates at −55 °C, where most CO_2 is separated, and the remaining CO_2 is captured by a hydrate phase operated at low temperatures of 1 °C [37]. This technology had the advantage of removing biogas impurities along with achieving CO_2 purity of 99%.

(d) *Cryogenic Absorption Process*

As shown in Fig. 9.6, this method integrates traditional MEA or NH_3-based biogas upgrading and the cryogenic process. This method provides the significant advantages of low costs and less energy required for regeneration [35].

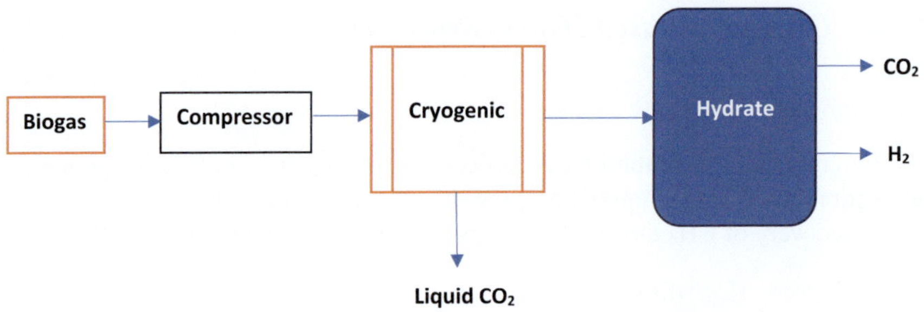

Fig. 9.5 Flow process of the cryogenic hydrate process. Adapted from [6, 7]

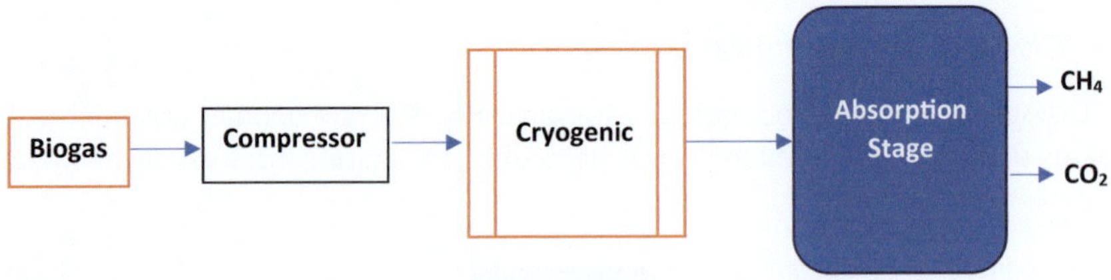

Fig. 9.6 Flow process of the cryogenic absorption process. Adapted from [6, 7]

9.5 Conclusion

Upgrading biogas is a field that has much room for innovation and advancement. To increase productivity and cost-effectiveness, there are chances to create new technologies and investigate various hybrid approaches—the fusion of established and cutting-edge technology. Cryogenic, biological, and hybrid technologies are recent advancements in biogas upgrading technology. There is very little data available because most procedures that are currently in use are still being developed. These techniques are still in the early stages of development and show promise, but they work better on lab scales. For biogas upgrading technology to be used more widely around the world, both established and new technologies are constantly being upgraded for higher performance, increased upgrading efficiency, and low cost.

References

1. Agneessens LM, Ottosen LDM, Voigt NV, Nielsen JL, de Jonge N, Fischer CH, Kofoed MVW (2017) In-situ biogas upgrading with pulse H_2 additions: the relevance of methanogen adaption and inorganic carbon level. Biores Technol 233:256–263. https://doi.org/10.1016/J.BIORTECH.2017.02.016
2. Alfaro N, Fdz-Polanco M, Fdz-Polanco F, Díaz I (2018) Evaluation of process performance, energy consumption and microbiota characterization in a ceramic membrane bioreactor for ex-situ biomethanation of H_2 and CO_2. Biores Technol 258:142–150. https://doi.org/10.1016/J.BIORTECH.2018.02.087
3. Anantharaman R, Berstad D, Roussanaly S (2014) Techno-economic performance of a hybrid membrane—Liquefaction process for post-combustion CO_2 capture. Energy Procedia 61:1244–1247. https://doi.org/10.1016/J.EGYPRO.2014.11.1068
4. Angelidaki I, Treu L, Tsapekos P, Luo G, Campanaro S, Wenzel H, Kougias PG (2018) Biogas upgrading and utilization: current status and perspectives. Biotechnol Adv 36(2):452–466. Elsevier Inc. https://doi.org/10.1016/j.biotechadv.2018.01.011
5. Aryal N, Kvist T, Ammam F, Pant D, Ottosen LDM (2018) An overview of microbial biogas enrichment. Biores Technol 264:359–369. https://doi.org/10.1016/J.BIORTECH.2018.06.013
6. Baena-Moreno FM, Rodríguez-Galán M, Vega F, Vilches LF, Navarrete B, Zhang Z (2019a) Biogas upgrading by cryogenic techniques. Environ Chem Lett 17(3):1251–1261. Springer Verlag. https://doi.org/10.1007/s10311-019-00872-2
7. Baena-Moreno FM, Rodríguez-Galán M, Vega F, Vilches LF, Navarrete B (2019b) Review: recent advances in biogas purifying technologies. Int J Green Energy 16(5):401–412. Taylor and Francis Inc. https://doi.org/10.1080/15435075.2019.1572610
8. Bailera M, Lisbona P, Romeo LM, Espatolero S (2017) Power to gas projects review: lab, pilot and demo plants for storing renewable energy and CO_2. Renew Sustain Energy Rev 69:292–312. https://doi.org/10.1016/J.RSER.2016.11.130
9. Bassani I, Kougias PG, Angelidaki I (2016) In-situ biogas upgrading in thermophilic granular UASB reactor: key factors affecting the hydrogen mass transfer rate. Biores Technol 221:485–491. https://doi.org/10.1016/J.BIORTECH.2016.09.083
10. Bassani I, Kougias PG, Treu L, Porté H, Campanaro S, Angelidaki I (2017) Optimization of hydrogen dispersion in thermophilic up-flow reactors for ex situ biogas upgrading. Biores Technol 234:310–319. https://doi.org/10.1016/J.BIORTECH.2017.03.055
11. Berg B, Clodic S (2017) Transforming biogas into Bio-LNG and liquid CO_2

12. Boontawee S, Koonaphapdeelert S (2016) In-situ biomethane enrichment by recirculation of biogas channel digester effluent using gas stripping column. Energy Procedia 89:78–84. https://doi.org/10.1016/J.EGYPRO.2016.05.011
13. Corbellini V, Kougias PG, Treu L, Bassani I, Malpei F, Angelidaki I (2018) Hybrid biogas upgrading in a two-stage thermophilic reactor. Energy Convers Manage 168:1–10. https://doi.org/10.1016/j.enconman.2018.04.074
14. Dolfing J, Jiang B, Henstra AM, Stams AJM, Plugge CM (2008) Syntrophic growth on formate: a new microbial niche in anoxic environments. Appl Environ Microbiol 74(19):6126–6131. https://doi.org/10.1128/AEM.01428-08
15. Facchino M, Popielak P, Panowski M, Wawrzyńczak D, Majchrzak-Kucęba I, De Falco M (2022) The environmental impacts of carbon capture utilization and storage on the electricity sector: a life cycle assessment comparison between Italy and Poland. Energies 15(18). https://doi.org/10.3390/en15186809
16. Goffeng B (2013) Crynotechnology for biogas. Lund University, Department of Chemistry Engineering
17. Johansson N (2008) Production of liquid biogas, LBG, with cryogenic and conventional upgrading technology-description of systems and evaluations of energy balances
18. Jonsson S, Westman J (2011) Cryogenic biogas upgrading using plate heat exchangers
19. Kapoor R, Ghosh P, Kumar M, Vijay VK (2019) Evaluation of biogas upgrading technologies and future perspectives: a review. Environ Sci Pollut Res 26(12):11631–11661. Springer. https://doi.org/10.1007/s11356-019-04767-1
20. Kougias PG, Treu L, Benavente DP, Boe K, Campanaro S, Angelidaki I (2017) Ex-situ biogas upgrading and enhancement in different reactor systems. Biores Technol 225:429–437. https://doi.org/10.1016/j.biortech.2016.11.124
21. Lecker B, Illi L, Lemmer A, Oechsner H (2017) Biological hydrogen methanation—A review. Biores Technol 245:1220–1228. https://doi.org/10.1016/J.BIORTECH.2017.08.176
22. Li Yuen Fong JC, Anderson CJ, Xiao G, Webley PA, Hoadley AFA (2016) Multi-objective optimisation of a hybrid vacuum swing adsorption and low-temperature post-combustion CO_2 capture. J Clean Prod 111:193–203. https://doi.org/10.1016/J.JCLEPRO.2015.08.033
23. Liu Y, Whitman WB (2008) Metabolic, phylogenetic, and ecological diversity of the methanogenic archaea. Ann N Y Acad Sci 1125:171–189. https://doi.org/10.1196/annals.1419.019
24. Luo G, Angelidaki I (2012) Co-digestion of manure and whey for in situ biogas upgrading by the addition of H_2: process performance and microbial insights. Appl Microbiol Biotechnol 97(3):1373–1381. https://doi.org/10.1007/s00253-012-4547-5
25. Luo G, Angelidaki I (2013) Hollow fiber membrane based H_2 diffusion for efficient in situ biogas upgrading in an anaerobic reactor. Appl Microbiol Biotechnol 97:3739–3744. https://doi.org/10.1007/s00253-013-4811-3
26. Luo G, Johansson S, Boe K, Xie L, Zhou Q, Angelidaki I (2012) Simultaneous hydrogen utilization and in situ biogas upgrading in an anaerobic reactor. Biotechnol Bioeng 109(4):1088–1094. https://doi.org/10.1002/bit.24360
27. Mulat DG, Mosbæk F, Ward AJ, Polag D, Greule M, Keppler F, Nielsen JL, Feilberg A (2017) Exogenous addition of H_2 for an in situ biogas upgrading through biological reduction of carbon dioxide into methane. Waste Manage 68:146–156. https://doi.org/10.1016/J.WASMAN.2017.05.054
28. Nishimura N, Kitaura S, Mimura A, Takahara Y (1992) Cultivation of thermophilic methanogen KN-15 on H_2-CO_2 under pressurized conditions. J Ferment Bioeng 73(6):477–480. https://doi.org/10.1016/0922-338X(92)90141-G
29. Pellegrini LA, De Guido G, Langé S (2018) Biogas to liquefied biomethane via cryogenic upgrading technologies. Renew Energy 124:75–83. https://doi.org/10.1016/J.RENENE.2017.08.007
30. Rachbauer L, Voitl G, Bochmann G, Fuchs W (2016) Biological biogas upgrading capacity of a hydrogenotrophic community in a trickle-bed reactor. Appl Energy 180:483–490. https://doi.org/10.1016/J.APENERGY.2016.07.109

31. Sahota S, Shah G, Ghosh P, Kapoor R, Sengupta S, Singh P, Vijay V, Sahay A, Vijay VK, Thakur IS (2018) Review of trends in biogas upgradation technologies and future perspectives. Bioresour Technol Rep 1:79–88. Elsevier Ltd. https://doi.org/10.1016/j.biteb.2018.01.002
32. Schmidt JE, Ahring BK (1993) Effects of hydrogen and formate on the degradation of propionate and butyrate in thermophilic granules from an upflow anaerobic sludge blanket reactor. Appl Environ Microbiol 59(8). https://journals.asm.org/journal/aem
33. Scholz M, Melin T, Wessling M (2013) Transforming biogas into biomethane using membrane technology. Renew Sustain Energy Rev 17:199–212. https://doi.org/10.1016/j.rser.2012.08.009
34. Shao P, Dal-Cin M, Kumar A, Li H, Singh DP (2012) Design and economics of a hybrid membrane-temperature swing adsorption process for upgrading biogas. J Membr Sci 413–414:17–28. https://doi.org/10.1016/j.memsci.2012.02.040
35. Song C, Liu Q, Deng S, Li H, Kitamura Y (2019) Cryogenic-based CO_2 capture technologies: state-of-the-art developments and current challenges. Renew Sustain Energy Rev 101:265–278. https://doi.org/10.1016/J.RSER.2018.11.018
36. Song C, Liu Q, Ji N, Deng S, Zhao J, Li Y, Kitamura Y (2017) Reducing the energy consumption of membrane-cryogenic hybrid CO_2 capture by process optimization. Energy 124:29–39. https://doi.org/10.1016/J.ENERGY.2017.02.054
37. Surovtseva D, Amin R, Barifcani A (2011) Design and operation of pilot plant for CO_2 capture from IGCC flue gases by combined cryogenic and hydrate method. Chem Eng Res Des 89(9):1752–1757. https://doi.org/10.1016/J.CHERD.2010.08.016
38. Tan Y, Nookuea W, Li H, Thorin E, Yan J (2017) Cryogenic technology for biogas upgrading combined with carbon capture-a review of systems and property impacts. Energy Procedia 142:3741–3746. https://doi.org/10.1016/j.egypro.2017.12.270
39. Tirunehe G, Norddahl B (2016) The influence of polymeric membrane gas spargers on hydrodynamics and mass transfer in bubble column bioreactors. Bioprocess Biosyst Eng 39(4):613–626. https://doi.org/10.1007/s00449-016-1543-7
40. Wang W, Xie L, Luo G, Zhou Q, Angelidaki I (2013) Performance and microbial community analysis of the anaerobic reactor with coke oven gas biomethanation and in situ biogas upgrading. Biores Technol 146:234–239. https://doi.org/10.1016/J.BIORTECH.2013.07.049
41. Yousef AM, El-Maghlany WM, Eldrainy YA, Attia A (2019) Upgrading biogas to biomethane and liquid CO_2: a novel cryogenic process. Fuel 251:611–628. https://doi.org/10.1016/J.FUEL.2019.03.127
42. Yousef AMI, Eldrainy YA, El-Maghlany WM, Attia A (2016) Upgrading biogas by a low-temperature CO_2 removal technique. Alex Eng J 55(2):1143–1150. https://doi.org/10.1016/J.AEJ.2016.03.026
43. Yun YM, Sung S, Kang S, Kim MS, Kim DH (2017) Enrichment of hydrogenotrophic methanogens by means of gas recycle and its application in biogas upgrading. Energy 135:294–302. https://doi.org/10.1016/J.ENERGY.2017.06.133

Chapter 10
Compressed Biogas (CBG) Landscape in India

10.1 Introduction

10.1.1 Historical Perspective of CBG—A Journey from the Past

In India, the history of biogas plants using cow dung (gobar gas plants) dates to 1859, when the first anaerobic digesters were installed in 1859 for biogas illumination. In 1946 Indian Agricultural Research Institute (IARI) developed a batch-type cattle dung digester, but its high cost hindered widespread adoption. Research continued to improve cost-effectiveness and gas production. The second phase (1950–1960) saw contributions from Jasbhai J. Patel and the IARI group. Patel designed the Gram Laxmi gas plant, patented in 1951, featuring a rotating gasholder and a continuous digestion system.

The third phase (1960 onwards) focused on the extension of biogas technology. Jasbhai J. Patel's Gram Laxmi III model improved upon previous designs with a brickwork digester and a gasholder. In the 1960s, the Gobar Gas Institute developed the KVIC model (Khadi and Village Industries Commission Model), and by 1973, over 6000 plants were commissioned.

In 1978, the Planning Research and Action Division (PRAD) developed the low-cost fixed dome Janta biogas plant, which was extended with the help of KVIC, AFPRO, and state agencies, installing 100,000 plants by 1980. Recognizing the significance of biogas technology, the Indian government launched the National Project on Biogas Development (NPBD) in 1981–82 to accelerate growth. In the 1980s, the Department of New Energy Sources (DNES) promoted biogas technology.

In 2002–03, the NPBD was renamed the National Biogas and Manure Management Programme (NBMMP). From 2005 to 06, biogas plants were disseminated through the Biogas-based Distributed/Grid Power Generation Programme and Energy Recovery from Urban and Industrial Waste Programs. In 2008–09,

V. K. Vijay et al., *Compressed Biogas (CBG)*, Green Energy and Technology,
https://doi.org/10.1007/978-981-96-5697-4_10

the government launched a new initiative for an Integrated Technology Package on medium-size biogas-fertilizer plants (BGFP), focusing on biogas generation, purification, bottling, and distribution.

Since 2014, India has shifted from traditional biogas to compressed biogas (CBG), driven by a focus on renewable energy and sustainability. Initiatives like the Sustainable Alternative Towards Affordable Transportation program (SATAT), Galvanizing Organic Bio-Agro Resources Dhan (GOBARdhan) scheme, and National Biofuel Policy (2018) and National Bioenergy Program (2023) have accelerated CBG adoption. State governments also implemented policies to encourage investment in CBG production. IIT Delhi has been instrumental in advancing CBG technology with cost-effective water scrubbing technology to enhance methane content, making it suitable for CBG production. Launched in 2018, the SATAT initiative aimed to set up 5000 CBG plants by 2024, producing 15 million metric tons of CBG annually. While progress has been slow, with only 113 plants commissioned by September 2025.

10.1.2 Forecast of India's Energy Demand

In the next two decades, India is projected to witness the most substantial increase in energy demand relative to any other nation, driven by its extensive development requirements. By 2050, India's proportion of world energy consumption will increase from 7 to 14%. Furthermore, India's oil and gas demand is projected to triple by 2050, with gas consumption expected to increase threefold by 2030. The International Energy Agency (IEA) predicts that India will account for 25% of the global energy demand increase from 2020 to 2040. The energy industry now generates around 75% of global greenhouse gas (GHG) emissions. It is vital in mitigating the most severe impacts of climate change, perhaps one of humanity's most critical crises [1].

10.1.3 India's Environmental Commitment and Role of CBG in Bioenergy Transition

In its "Nationally Determined Contribution" (NDC) to the "United Nations Framework Convention on Climate Change" (UNFCCC), India pledged to reduce its GDP's emissions intensity by 45% from 2005 levels by 2030. India is committed to achieving net-zero emissions of greenhouse gases by 2070 and plans to generate almost half of its overall electrical output from renewable sources by 2030. The Glasgow Climate Pact, agreed upon at the 26th Conference of the Parties (COP 26) in November 2021 in Glasgow, Scotland, aimed at turning the 2020s as a decade of environmental engagement and climate mitigation. Enhancing the shift to a gas-based economy and prioritizing indigenous sources to develop biofuels were two of the seven important parts of India's energy plan that the prime minister outlined at the conference [2].

Bioenergy comes from plant or other biological waste-derived organic materials. One of the most important ways to combat climate change is modern bioenergy, which uses cutting-edge technology to produce fuels from biomass in the form of solids, liquids, and gases for commercial use. It helps make communities around the world more resilient by lowering emissions of greenhouse gases, increasing the reliability of regional energy supply, bringing in money for rural areas, and making energy systems more adaptable. Compressed biogas (CBG), compatible with the current gas infrastructure, offers the most straightforward pathway to decarbonizing natural gas applications. CBG reduces energy system management costs because the infrastructure for transporting and using it exists. Regionally, "fermented organic manure" from anaerobic digestion can be utilized as a sustainable biological fertilizer after upgrading or alone [2].

The CO_2 extracted from biogas to produce CBG can also be regarded as a co-product. Biogas systems serve multiple functions beyond the substitution of fossil fuels. In addition to reducing pollutants, they improve waste handling and promote a business model that is circular. Moreover, they have the advantage of local installation and operation, ensuring a reliable energy supply and enhancing energy security. Generating biogas and "compressed biogas (CBG)" directly influences and bolsters multiple "*Sustainable Development Goals (SDGs)*," which are 17 worldwide objectives instituted by the United Nations General Assembly in 2015, with a target for completion by 2030. Their connection (mainly 15) is shown in Fig. 10.1 [2].

10.1.4 Installed Capacity of CBG Plants Across Indian States

As per the GOBARdhan portal, Ministry of Jal Shakti, 792 CBG plants have been registered in India, of which 112 are functional, 195 are under construction, and 485 are yet to start construction. Uttar Pradesh (U.P), Maharashtra, Punjab, Madhya Pradesh (MP), Haryana, and Gujarat lead the pack in the maximum registration of CBG plants. At the same time, states in the country's northeastern region, J&K and Himachal Pradesh, have the least registered plants. Similarly, states like the U.P. and Maharashtra have a maximum no. of functional plants across India. A complete list of functional and non-functional registered CBG plants across the different states is shown in Table 10.1, and a graphical presentation of registered CBG plants in different states is shown in Fig. 10.2 [3].

10.1.5 CBG Potential of India

CBG is an enriched form of biogas compressed to a high pressure of 200 bar with removed impurities like CO_2, H_2S, and moisture. The resultant gas has at least

Relevant SDG	CBG contribution	Relevant SDG	CBG contribution
1 NO POVERTY	Helping small-scale farmers by providing them with affordable fertilizer and addressing the challenges associated with the complicated process of getting fertilizer to them.	9 INDUSTRY, INNOVATION AND INFRASTRUCTURE	Establishing a sustainable infrastructure and creating value from waste through the production of biogas.
2 ZERO HUNGER	Boosting the efficiency of farming by improving the quality of soil through the restoration of depleted nutrients, organic materials and carbon.	11 SUSTAINABLE CITIES AND COMMUNITIES	Preventing diseases by collecting organic waste and treating wastewater, while simultaneously decreasing unpleasant odours and improving air quality, as well as increasing access to green energy.
3 GOOD HEALTH AND WELL-BEING	Decreasing the level of contact with both methane gas discharges and chemical fertilizers.	12 RESPONSIBLE CONSUMPTION AND PRODUCTION	Improving the utilization efficiency of natural resources by decreasing air and water pollution, and enhancing the process of waste recycling.
4 QUALITY EDUCATION	Enhancing the availability of energy in rural regions would enhance the standard of education.	13 CLIMATE ACTION	Lowering GHG emissions by offering an energy source that emits less, as well as minimizing methane emissions from various industries.
5 GENDER EQUALITY	Supplies an inexpensive energy source to the nearby villages in rural regions, thereby enhancing the standard of living for women and children.	14 LIFE BELOW WATER	Preventing pollution from land sources as a means of decreasing marine pollution.
6 CLEAN WATER AND SANITATION	The availability of energy could lead to an increase in wastewater treatment capacity, which could improve water quality. Additionally, managing household and animal waste may also enhance hygiene conditions.	15 LIFE ON LAND	Decreasing deforestation through the substitution of solid fossil fuels with biogas, while also improving land-use productivity and minimizing land-use change.
7 AFFORDABLE AND CLEAN ENERGY	Enhancing energy sustainability through improved reliability and affordability of energy supply, as well as increased energy storage capacity.	16 PEACE, JUSTICE AND STRONG INSTITUTIONS	Improved energy access is related to peace. It can contribute to economic development and job creation, reducing poverty and social inequality, which are often root causes of conflict
8 DECENT WORK AND ECONOMIC GROWTH	Improving waste management practices and reducing the carbon footprint of materials can boost the GDP.		

Fig. 10.1 CBG contribution to the concerned SDG [2]

Table 10.1 List of functional/non-functional registered CBG plants across states in India as of January 2025 [3]

State/UTs	Non-functional	Functional	Total
A & N islands	0	0	0
Andhra Pradesh	27	5	32
Arunachal Pradesh	0	0	0
Assam	4	0	4
Bihar	24	2	26
Chandigarh	1	0	1
Chhattisgarh	17	1	18
D & N Haveli and Daman & Diu	0	0	0
Delhi	5	0	5
Goa	0	0	0
Gujarat	36	18	54
Haryana	45	12	57
Himachal Pradesh	1	0	1
Jammu & Kashmir	1	0	1
Jharkhand	6	0	6
Karnataka	23	7	30
Kerala	5	0	5
Ladakh	0	0	0
Lakshadweep	0	0	0
Madhya Pradesh	51	5	56
Maharashtra	104	10	114
Manipur	0	0	0
Meghalaya	0	0	0
Mizoram	0	0	0
Nagaland	0	0	0
Odisha	17	0	17
Puducherry	0	0	0
Punjab	69	9	78
Rajasthan	24	6	30
Sikkim	0	0	0
Tamil Nadu	34	8	42
Telangana	23	4	27
Tripura	0	0	0
Uttar Pradesh	141	21	162
Uttarakhand	10	2	12
West Bengal	12	2	14
Total	**680**	**112**	**792**

Bold indicates the total no. of functional/non-functional registered CBG plants across the different states of the country i.e. India

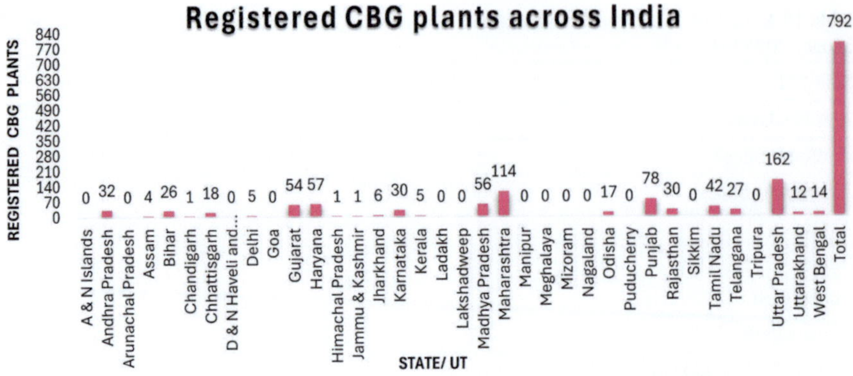

Fig. 10.2 Graphical presentation of registered CBG plants across India as of January 2025 [3]

Table 10.2 Composition of CBG [2]

CBG composition	Values
"Methane (CH_4)"	Greater than 90%
"Carbon Dioxide ($C0_2$)"	Less than 4%
"Nitrogen"	Less than 0.5%
"Oxygen"	Less than 0.5%
"Hydrogen Sulphide (H_2S)"	Less than 16 ppm
"Moisture content"	Nil
"CV"	46–53 MJ/kg

90% methane and is used as a vehicular fuel substitute for CNG and diesel. Thus, composition and properties mark the critical difference between CBG and raw biogas, as shown in Table 10.2.

India's total CBG generation potential is approximately 62 MMT from various sources, while less than 1% is utilized in installed capacity. Thus, our country has enormous potential to grow in this sector. Compressed biogas' estimated generation potential per different resources and contribution has been shown in Table 10.3 and Fig. 10.3, respectively.

Table 10.3 Estimate CBG generation potential and contribution per different resources [2]

Categories of organic waste	Estimated CBG generation potential (MMT)
Waste from animals	25
Surplus agro waste	20
STP "*Sewage Treatment Plants*" waste	10
Municipal solid waste (MSW)	5
Press mud, etc.	2

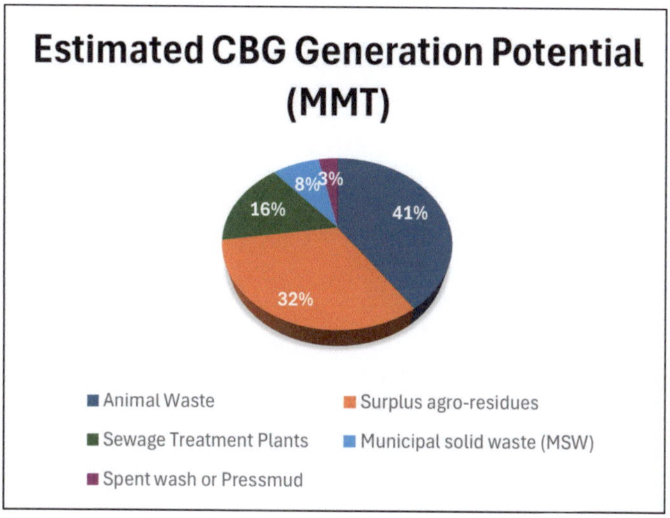

Fig. 10.3 Graphical presentation of contribution of estimated CBG generation potential per different resources [2]

Considering this, the Indian government announced a budget of Rs. 35,000 crores on February 1, 2023, to enhance energy security and hasten the shift to sustainable energy. The "*Galvanizing Organic Bio-Agro Resources Dhan* (*GOBARdhan*)" program will get a sizeable proportion of this budget, Rs. 10,000 crores. Through the installation of 500 new waste-to-wealth bio-CNG facilities, compressed biogas will be produced from organic waste. "*Out of the 500 bio-CNG plants, 200 will be CBG plants, and 75 will be in urban areas. The remaining 300 plants will be community or cluster-based. Additionally, the government has introduced the CBG blending obligation (CBO). CBO shall be kept as 1%, 3% and 4% of total CNG/PNG consumption for FY 2025–26, 2026–27 and 2027–28 respectively. From 2028–29 onwards CBO will be 5%.*

10.2 CBG Supportive Policies

The Government of India has launched several central and state policies to bolster the CBG sector's growth in the country owing to its energy security needs and global environmental commitments, as discussed in further sections.

10.2.1 Central Level

(a) **SATAT Scheme: "Sustainable Alternative towards Affordable Transportation"**

Launch: October 1, 2018.

Concerned Ministry: "Ministry of Petroleum and Natural Gas (MoPNG)".

Support *Area*: Gas offtake.

Key Highlights:

With the help of 5000 plants spread throughout India, this program sought to produce 15 million metric tons of CBG and 50 million metric tons of biomanure. The project aimed to produce compressed biogas (CBG) from biomass and waste sources, including sewage treatment plant waste, cow manure, sugarcane press mud, agricultural residue, and municipal solid waste (MSW). It guaranteed CBG that oil and gas marketing companies would purchase from the plants at a certain price. The procurement price of CBG was tied to the current retail selling price of CNG in the market, and the minimum procurement price, including applicable taxes, was set at Rs. 54 per kg. Addressing urban air pollution (NCR), processing and disposing of municipal solid waste, improving India's energy security, encouraging entrepreneurship and the rural economy, and making modern energy easily accessible and affordable are just a few advantages of this project. As of November 2024, 77 plants have been commissioned with 2237 active LOIs. Besides, 17,801 tons of CBG have been sold till November, 2024 [4].

Further Details: "https://satat.co.in/satat/#/about".

(b) **MNRE: "Waste to Energy Scheme"**

Launch: November 2, 2022.

Concerned Ministry: "The Ministry of New and Renewable Energy (MNRE)".

Support Area: Incentive on plant capital expenditures.

Key Highlights:

The program aims to make setting up waste-to-energy plants that turn agricultural, industrial, and urban trash and residues into CBG easier. With a maximum CFA of Rs. 10.0 Cr/project, CBG project developers are eligible to earn Rs. 4 crores for new CBG plants and Rs. 3 crores for existing plants with a capacity of 4800 kg/day of CBG output for both the cases. Prior to the waste-to-energy (WTE) plant being put into service, the proposal for the award of "In-Principle" approval of Central Financial Assistance (CFA) will be accepted via the "BioURJA Portal (https://biourja.mnre.gov.in)" [5].

Further Details: "https://mnre.gov.in/en/waste-to-energy/".

(c) **Market Development Assistance**

Launch: September 20, 2023.

Concerned Ministry: "Ministry of Chemicals and Fertilizers".

Support Area: Incentive for fermented organic manure (FOM).

Key Highlights:

With a budget allocation of Rs. 1451.82 crore for three years (FY 2023–24 to FY 2025–26), this project was created to promote the production of fermented organic manure (FOM), liquid fermented organic manure (LFOM), and phosphate-rich organic manure (PROM) in biogas/CBG facilities. The CBG producers will get MDA of the FOM/LFOM/PROM at Rs. 1500 per ton. To receive MDA, the CBG unit must first register on the Department of Drinking Water and Sanitation's GOBARdhan portal (Cabinet Committee on Economic Affairs [6].

(d) **Biomass Aggregation Machinery (BAM)**

Launch: February 2, 2024.

Concerned Ministry: "Ministry of Petroleum and Natural Gas (MoPNG)".

Support Area: Incentive on *"BAM"*.

Key Highlights:

By giving CBG producers financial help to buy "Biomass Aggregation Machinery (BAM)," this program seeks to facilitate biomass collection for the first 100 biomass-based CBG facilities. The maximum financial aid offered under this program is Rs. 90 lakhs per set, or 50% of the cost of purchasing BAM, whichever is lower. The support is available to all current and planned CBG plants with an established CBG production capacity of at least 2 TPD and utilizing at least 50% biomass (agri residue) as feedstock [7].

Further details for eligibility criteria: "https://bam.eil.co.in/Default".

(e) **Development of Pipeline Infrastructure (DPI)**

Launch: 26 June 2024.

Concerned Ministry: "Ministry of Petroleum and Natural Gas (MoPNG)".

Support Area: Incentive on pipeline infrastructure.

Key Highlights:

The GoI announced this proposal to provide funding for the construction of a pipeline that would link the CBG facility to the current CGD network. FA for a maximum of 50 km would be offered at Rs. 0.5 crore per kilometer, or 50% of the project cost, whichever is lower. The FA will be limited to 50% of the project cost or 0.075 crores per kilometer, whichever is less, in the event of connection when only the MDPE pipeline is utilized. FA would permit up to 75 km of pipes. (MDPE/Steel) The FA pipeline's maximum cost per project will be Rs. 28.75 crore.t [8].

Further details for eligibility criteria: "https://dpi.eil.co.in/Default".

(f) **Other Financial Assistance Frameworks**

Apart from the above policies, financial assistance to the CBG sector is also provided in ways like the "Agricultural Infrastructure Fund" (AIF), which offers funding for investments in feasible initiatives, including community farming assets and postharvest management infrastructure. Up to a loan amount ceiling of Rs. 2 crores, all loans made possible by this financing facility would be given an interest subvention of 3% annually for a maximum of seven years. It is monitored by the *"Ministry of Agriculture and Farmers' Welfare"* [2].

CBG is a clean and renewable fuel, thus by producing and utilizing it, facilities can receive carbon credits for lowering greenhouse gas emissions. For CBG facilities, the potential earnings of carbon credits may offer an extra revenue stream. To encourage sustainable growth and lower greenhouse gas emissions, several Indian businesses have recently allocated cash under CSR vertical to CBG initiatives. With *"Mahindra & Mahindra Ltd.'s"* CSR funding, Mahindra Waste to Energy Solutions has established CBG facilities in Pune and Mumbai. Besides, priority sector lending established by RBI can also be used for sector growth [2].

10.2.2 State Level

With only 100 CBG projects operational as of August 2024, India's adoption of CBG projects has been sluggish and well behind the target of 5000 plants by 2030. The inefficient implementation of state-level regulations is a major barrier to the growth of CBG. Despite having substantial CBG potential, most state governments lack well-defined incentives and legislation. As of 2024, only for states with bioenergy policies encourage CBG: Gujarat, Haryana, Bihar, and Uttar Pradesh. These four states can act as a model for other governments looking to enter the market. The bioenergy policies of the states are contrasted in Table 10.4. The best state for comprehensive support in catchment regions, land aid, and subsidies is Uttar Pradesh. The maximum subsidy amount is Rs. 75 lakhs per ton of CBG in addition to the central government

Table 10.4 Comparison of state-based bioenergy policy [9]

Support vertical	Bihar	Gujarat	Uttar Pradesh	Haryana
Capital assistance	Up to 15% of the cost of plant and machinery (max. crores)	25% of eligible Gross Fixed Capital Investment	75 lakhs/ton of CBG production (max. 20 crore)	N/A
Electricity duty	N/A	15% of cost incurred (max. 15 crore)	N/A	N/A
Land A assistance	N/A	Yes, for Identification	On lease available for Rs1/acre/year	N/A
Catchment area	N/A	N/A	One plant per Tehsil	N/A
Online portal	(https://swc2.bihar.gov.in)	(https://btm.gujarat.gov.in/bt-policy-new.htm)	(https://www.upneda.in/bio/)	No
Duration of scheme	10/07/ 2023–30/06/2025	15/02/2022–31/03/2027 2027	10/ 2022–10/2027 2027	March 2018–not defined

subsidy. In conclusion, states promote CBG to varying degrees, but the scope and character of these advantages differ greatly [9].

10.3 Regulatory Approvals for CBG Plant Setup

Regulatory approvals must be taken from government agencies to set up a CBG plant in India. A consolidated list of such approvals is shown in Table 10.5.

Table 10.5 List of CBG plant setup regulatory approvals [2]

S. No.	Type of approval	Pre-establishment	Post-establishment (pre-operation)	Yearly renewal
1	Non-industrial land to non-agricultural land	Yes, with registration	No	No
2	Fire department	No objection certificate	License required	Yes
3	Indian ordnance factories plan	Building and construction workers registration and NOC required	Subcontractor license required	Yes
4	Health and safety department	No objection certificate	License required	Yes
5	District industry center	MSME registration required	No	No
6	District town and country planning	No objection certificate	Running license required	No
7	Block development officer—Running license	No	Yes	Yes
8	Local panchayat	NOC, tax payment, and plan submission based on construction area	Annual tax, if applicable	No
9	Petroleum and explosives	Consent for construction required	Consent for operation required	Yes
10	Pollution control board	Consent for construction required	Consent for operation required	Yes

10.4 Role of GIS in CBG Plant Installation—An ICT-Driven Approach

By facilitating the effective selection of sites through the study of biomass availability, land use, water resources, and accessibility to utility and transportation networks, GIS is essential to the establishment of compressed biogas (CBG) facilities. Feedstock movement cost reduction and transportation route planning aid in logistics optimization. By analyzing topography, population density, and ecological sensitivity, GIS also aids in environmental evaluations and infrastructure development. It also facilitates the use of remote sensing to track trends in biomass output and operational performance. GIS improves cost-effectiveness, sustainability, and decision-making in the construction and management of CBG facilities. It is the need of the hour that CBG optimal site selection should be done with the help of GIS across the country to effectively plan the growth of this sector in the upcoming two decades for energy self-reliance.

Besides, many factors must be carefully considered when choosing a CBG plant location to guarantee the plant's safe and effective operation. These are some important factors to consider when choosing a CBG facility location.

(1) **Consistent and Dependable Feedstock Supply**

The manufacturing of CBG necessitates a certain kind of feedstock, which falls into four primary categories: municipal, industrial, animal, and agricultural waste. There are fewer contaminants and a more uniform quality in agricultural waste. It is not economically viable to carry agricultural feedstock and stackable energy crops further than 15 km because to their huge amount and low energy content per unit of volume. Therefore, it is advised that a biogas plant be built within 15 km of the easily accessible biomass. The total amount of agricultural residue available within this 15-km radius should be four times the actual plant's need to guarantee feedstock sustainability measures. Long-term agreements with farmer-producer associations or secondary biomass suppliers should be used to secure at least 80% of the feedstock supply [2].

Care must be used while defining the quantity and quality of supply obligations. When establishing a 3.5-ton capacity CBG plant, Ather Mehmood, an investor in these types of plants in Bulandshahr in Uttar Pradesh, highlighted feedstock availability as the most important factor. He selected a site where various feedstocks, including press mud, cow manure, and agricultural waste, were easily accessible within a 15-km radius. Additionally, he investigated the potential for growing Napier grass nearby. A minimum of 30 km must be kept between CBG facilities or other comparable setups that use the same biomass feedstock in an area to avoid market disputes and price swings brought on by conflicting interests for agricultural feedstock [2].

(2) **Commercial Viability and Product Offtake Network**

Organic fertilizers and compressed natural gas are the two main outputs of the CBG facility. A comprehensive analysis of a 15–20 km radius is required to guarantee the commercial viability of CBG, considering variables like the quantity of CNG pumps, the accessibility of the gas supply network, the companies that depend on CNG as fuel, and the estimated number of cars that have been converted to run on CNG. To guarantee a seamless transaction and establish a fair price for the goods, the overall regional demand must be greater than the production capacity. Digestate, the plant's other important product, must also be promoted and sold. In addition to being a great soil conditioner, it may be enhanced with nutrients suited to a certain crop to provide high-value organic fertilizer. To assess the market size for biogas digestate, market research must be done to find potential consumers, such as greenhouses, residential landscaping, and agricultural land cultivating organic products [2].

(3) **Land Use Pattern and Availability**

Installing a CBG plant benefits from having uncontested and clear land title as it facilitates bank and other investor financing. Additionally, this increases the project's credibility with investors and removes the possibility of delays brought on by obtaining agreements from other parties. Furthermore, since the biogas plant needs a steady supply of biomass, which causes a large outflow of biomass, access to sturdy roadways is essential. To preserve the landscape, industrial sites should be prioritized above green spaces when choosing a location for the plant. Soil stability must also be guaranteed to prevent further expenditures on enhancing subterranean conditions. A separate feeder connection from a nearby power substation should be investigated if the facility needs electricity around the clock because of automation. The plant's unavoidable noise and odor emissions must also be considered. Therefore, selecting an appropriate environment that will reduce future disputes is crucial [2].

(4) **State Peculiar Advantages**

Several states offer special incentives to encourage the installation of CBG plants inside their borders. These states have bioenergy policies, which facilitate and expedite the regulatory clearance procedure for installing CBG plants. In 2022, for example, Uttar Pradesh unveiled a new bioenergy strategy that incentivizes Rs. 75 lakh per ton of compressed biogas output. Benefits, including tax breaks, free public land, and agriculture equipment subsidies, are also included in the strategy. An online bioenergy portal has also been set up to promote openness and expedite the single-window approval procedure [2].

(5) **Acquiring Institutional and Political Endorsement**

Because of the high level of community and press curiosity in renewable energy, building a bio-CNG plant is politically complex and needs strong public support. A consistent and deliberate communication effort must be conducted to highlight the project's economic and environmental benefits to the area to secure institutional and political support. Organizing visits to CBG sites through cooperative tours with central or regional institutes has been beneficial, presenting the plant as a "Best Practice" example for waste management or sustainable energy generation [2].

10.5 Case Studies

10.5.1 Verbio AG CBG Plant, Sangrur, Punjab

A 33 TPD of CBG production capacity, eight-digester CBG plant was established on June 2022. Paddy straw and other agricultural waste are used for feedstock, while chemical absorption is used for biogas enrichment. Gas off-take is done via cascades to nearby CNG pumps. The CAPEX cost of the plant is around Rs. 230 crore.

This project, which aims to improve farmers' livelihoods and reduce air pollution, has yielded enormous advantages. With an annual consumption of one lakh tons of agricultural leftovers, the operation has considerably reduced air pollution from stubble burning within a minimum 10–15 km radius. Furthermore, farmers save significant money because the entire stubble-collecting operation is done for free. A privately owned and administered agro-residue supply and distribution system that combines infrastructure and procedures for feedstock accumulation, distribution, storage, and control has been established because of the project. Consequently, it has boosted economic growth and produced jobs in rural areas. Additionally, the project's delivery of biomanure to nearby fields has given farmers a genuine chance to improve soil quality [2].

10.5.2 Mittal Enterprises, Garh Mukteshwar, Uttar Pradesh

An 8 TPD of CBG production capacity, two-digester CBG plant was established on January 1, 2023. Pressmud, i.e., sugarcane industrial waste, is used for feedstock, while water scrubbing technology is used for biogas enrichment. Gas off-take is done via cascades per the contract with Indraprastha Gas Limited. The CAPEX cost of the plant is around Rs. 26 crore.

Situated next to five sugar manufacturing facilities, the CBG factory uses copious amounts of pressmud produced during sugarcane processing. Sugar production efficiently decreases waste by generating enough pressmud to meet the plant's fuel needs. Since the debris from the sugarcane processing business is frequently burnt or disposed of in landfills, the environmental effects of this activity have long been a source of worry. A CBG facility that transforms this waste into clean fuel has been established to lessen its environmental effects. One unique feature of this facility is its utilization of cutting-edge water scrubbing technology to convert biogas into premium CBG [2].

10.5.3 EverEnviro Resource Management Private Limited, Indore, MP

A 17–18 TPD of CBG production capacity, a four-digester CBG plant was established on February 19, 2022. Biodegradable municipal waste is used for feedstock, while the pressure swing adsorption method is used for biogas enrichment. Gas off-take is done to IMC city buses and industrial or commercial use. The CAPEX cost of the plant is around Rs. 150 crore.

With the Indore Municipal Corporation (IMC) collaborating with EverEnviro to make this project a reality, the Indore plant is proof of the effectiveness of public–private partnerships. EverEnviro has made it feasible to transform the garbage that formerly plagued Devguradiya into a lucrative product. In exchange for EverEnviro paying them an annual royalty of Rs. 2.5 crore, the IMC provides the plant with 90% of the separated garbage. According to the terms of the agreement between the IMC and EverEnviro, half of the CBG generated shall be used to power city buses in Indore, priced at Rs. 5 per kilogram below the going rate for CNG. The Indore region's CNG and PNG provider, Avantika Gas Ltd., purchases the remaining CBG. One of the main reasons for the Indore plant's success is the city's remarkable trash segregation at the point of origin. The cost of producing CBG is as low as Rs. 35–40 per kg, demonstrating the success of IMC's efforts, By demonstrating that it is feasible to transform garbage into treasure, the Indore plant's creative approach to waste management serves as a model that other communities may be inspired to adopt [2].

10.5.4 Hingonia Gaushala, IOCL, Jaipur, Rajasthan

With a 6 TPD of CBG production capacity and two digesters, CBG plant was established in January 2023. Cattle dung is used for feedstock, while pressure swing adsorption technology is used for biogas enrichment. Gas off-take is done in the Akshaya Patra Foundation's kitchen to prepare meals. The CAPEX cost of the plant is around Rs. 38 crore and was built using Indian Oil Corporation Ltd.'s CSR funds.

Interestingly, using this gas, 150,000 school children's lunches are prepared daily by the Akshaya Patra Foundation. This well-known nonprofit provides school children all around India with midday meals. The foundation is an excellent example of social accountability and sustainability. This initiative benefits the environment by lowering greenhouse gas emissions and reducing reliance on non-renewable resources [2].

10.5.5 IIT Delhi Biogas Plant, Gramodaya Parisar, IIT Delhi Campus

A 25 m^3/day CBG production capacity, one-digester plant was established in 2004 at Gramodaya Parisar, a 7-acre piece of land at the IIT Delhi campus, Hauz Khas, as shown in Fig. 10.4. Kitchen waste collected from over 200 households within the IIT Delhi campus is used as feedstock. In contrast, water scrubbing technology (indigenously developed and patented) is used for biogas enrichment and bottling. Gas off-take is done for research purposes for both the 2-wheeler and 4-wheeler. The capital expenditure cost of the plant is around Rs. 40 Lakhs. The word "CBG" was first coined by Prof. VK Vijay and his research group at the Biogas Lab in the Centre for Rural Development and Technology, sponsored by MNRE in 2008, which was later published in the IIT Delhi Golden Jubilee Calendar in 2011, as shown in Fig. 10.5.

Interestingly, the standard on biomethane, IS 16087, has been established by the data collected from a WagonR car, sponsored by MNRE and driven by Prof VK Vijay over the last 12 years, covering more than 1,25,000 km, as shown in Fig. 10.6. One can find the details on the link: https://www.bis.gov.in/. Recently, BIS standards for CBG have been updated in 2025. A comparative analysis of BIS standards for CBG in 2016 vs 2025 is depicted in the below Table 10.6. Research is ongoing on CBG-based 2-wheelers and tractors, as shown in Figs. 10.7 and 10.8. Besides, IIT Delhi has also developed its own Biogas Enrichment and Bottling Technology (water scrubbing), which is patented and transferred to more than nine industries, as shown in Fig. 10.9.

Table 10.6 Comparative Analysis of BIS Standards for CBG 2016 vs 2025 [20]

Parameter	IS 16087:2016	IS 16087:2025	Important Updates
Methane (CH_4)	$\geq$ 90% v/v	$\geq$ 95 mol%	Purity requirements made stricter
$CO_2 + N_2 + O_2$	$\leq$ 10% v/v	$\leq$ 5 mol%	Impurities reduced by half
CO_2 (only)	$\leq$ 4% v/v	$\leq$ 4% mol	Same limit retained
O_2 (only)	$\leq$ 0.5% v/v	$\leq$ 0.5% mol	No change
Total Sulphur (incl. H_2S)	$\leq$ 20 mg/m^3	$\leq$ 10 mg/m^3	Limit tightened by 50%
H_2S	Not separately specified (part of total sulphur)	$\leq$ 3.7 mg/m^3 (new dedicated limit)	New specific limit introduced
Moisture	$\leq$ 5 mg/m^3 (Karl Fischer titration)	$\leq$ 5 mg/m^3 (Karl Fischer titration + coulometric)	An additional measurement method was introduced
Cylinder Storage	IS 7285 (Pt. 2) & IS 15490 (steel cylinders)	IS 7285 (Pt. 2), IS 15490, plus IS 15935 (composite cylinders)	Composite cylinders permitted
Marking & Certification	BIS Act 1986	BIS Act 2016	Updated legal framework
Regulatory Alignment	MoRTH focus (automotive)	PNGRB focus (grid blending with CGD)	Application scope expanded

Fig. 10.4 Kitchen waste-based 25 m^3/day biogas plant at Mahatma Gandhi Gramodaya Parisar, CRDT, IIT Delhi

Fig. 10.5 IIT Delhi golden jubilee calendar, 2011

Fig. 10.6 Biogas car sponsored by MNRE, IIT Delhi

Fig. 10.7 CBG-driven two-wheeler at IIT Delhi

Fig. 10.8 Dual fuel mode (diesel and CBG) tractor at IIT Delhi

Fig. 10.9 Biogas enrichment technology (water scrubbing) developed at IIT Delhi

10.6 Obstacles/Risks—The Industry Faces

1. **No Purchase of Biogas Generation-Based Organic Fertilizer**

A sizable amount of the by-product is the residue—also referred to as digestate or slurry—created during anaerobic digestion. Due to a lack of set procurement costs, legal restrictions, and low fertilizer company demand, digestate is not completely utilized. It is a rich source of carbon and nitrogen that may improve soil health by enhancing soil structure and boosting water retention [2].

2. **Unsupervised Supply Chain Management of Biomass**

The steady supply of high-quality raw materials significantly impacts a CBG facility's profitability and efficacy. However, the one-and-a-half month window for collecting agricultural feedstock and the volatility of feedstock prices make it difficult to keep the plant financially viable. Harvesting agricultural residue is accomplished by farm equipment such as tractors, rakers, balers, loaders, and trailers. There is a lack of equipment to harvest and move the feedstock to biogas facilities. Furthermore, CBG investors cannot gather this enormous trash independently [2]. To sum up, huge **operational risk** looms across the sector in the country.

3. **Financing CBG Projects at a Huge Cost**

A substantial amount of cash is needed to establish a CBG facility. Banks charge exorbitant interest rates for loans pertaining to CBG, even though the Reserve Bank of India has designated this industry as a priority area for financing. Loans for such ventures can only be obtained from a select group of private banking institutions [2]. Thus, these **financial risks** in combination with higher capital costs and insurance gaps act as major barriers to fully utilize the CBG potential in our country.

4. **Ineffective Feedstock Source Segregation**

Biogas facilities have challenges due to improper source segregation of waste from municipalities, which may pollute the feedstock used to produce biogas. Improper trash sorting can result in non-biodegradable items like metals and plastics, harming the biogas plant's machinery and lowering the method's effectiveness. Furthermore, producing high-quality biogas from organic waste in conjunction with non-organic waste would be improper, impacting the biogas plant's total production [2]. This, in combination with seasonal availability of feedstock and volatile prices of feedstock in competing biomass demand, poses a major **resource risk** to the CBG sector in our country.

5. **Insufficient Manufacturing of Indigenous Plant Equipment**

Much of the gear and equipment needed for CBG facilities, including digester material, biogas enrichment units, storage vessels, compressors, dispensers, and technology or equipment for monitoring and control, are frequently imported and may not always be readily available. This can make it difficult to set up and run CBG operations, especially in places like India, where the local manufacturing sector for such machinery is lacking. Additionally, importing these parts might result in higher prices, more delays, and even supply chain interruptions [2]. Additionally, unoptimized CBG plant operating efficiency and the use of conventional technologies act as major **technological risks** for the sector in India.

6. **Absence of CNG Users in Rural and Distant Regions**

The dearth of CNG users in rural regions is one of the obstacles to the adoption of CBG. Rural locations may not have enough demand for CNG as a transportation fuel because of their lower population density and limited supply of CNG cars, in contrast to metropolitan areas with a significant demand. Because the expenses of making and delivering CNG could not be covered by the money made from sales, this could restrict the economic feasibility of CBG establishments in rural regions [2].

10.7 Sustainable Implementation of CBG Plants in India and Way Ahead

Indian Institute of Technology Delhi (IITD) has been working on biogas production, enrichment, and bottling technologies for over 25 years. A significant contribution has been made to biogas enrichment and bottling for vehicular applications. It started by evaluating all contemporary technologies, such as water scrubbing technology, cryogenic separation, pressure swing adsorption, and membrane separation, available worldwide for biogas upgradation suitable to Indian conditions. Considering a rural base of 700,000 villages in India and having the largest cattle population, it has been evaluated that water scrubbing technology is a more suitable option considering environment friendliness, local availability of resources, skill set requirements of villages, low repair and maintenance cost, availability of water, and less hazardous chemicals. All other technologies require chemicals, zeolite, or a membrane, which are imported from other countries, and it is observed that slowly, their effectiveness (efficiency) decreases as time passes.

After evaluating these parameters of the biogas upgradation system, IIT Delhi finalized the development of a low-cost, small-scale water scrubbing system. After a PhD work in water scrubbing technology for biogas enrichment and successful experimentation, the development was patented in 2006 and licensed to many industries and *goshalas* (large cattle sheds). **Then, IITD coined a new name for this technology, compressed biogas (CBG), rather than calling it biomethane or Bio-CNG,** which were popular terms in other global parts. Apart from biogas systems and their regular and transport applications, a time has come when biogas is seen as a green, environment-friendly, and carbon–neutral technology, looking after climate change mitigation and reducing greenhouse gas emissions. It can provide many jobs in rural areas by utilizing many types of available organic waste for biogas generation.

In India, more than 50% of the village population depends on agricultural and animal husbandry activities for their livelihood. CBG technology can also provide solutions for waste management and harnessing waste for energy and biofertilizer applications. If a cluster of 7–9 villages is taken, aggregating all kinds of organic waste material such as cow dung, fecal waste, agricultural residues, kitchen waste, rotten fruits, vegetables, etc., and other available organic industrial wastes available in surplus in the region in a 10–15 km radius, then it can be utilized for biogas production. It will ensure a good supply chain for successful biogas production. Therefore, there is a need to develop multi-feed digesters that can take care of all kinds of locally available wastes. In addition, there is a need to bolster and develop a supply chain management system across the country for feedstock such as paddy straw, press mud from sugar mills, waste from food processing industries, and MSW in cities and urban landscapes.

Thus, in line with the existing policy of large-scale CBG plants by GoI, there is a need to develop a parallel ecosystem of community-based CBG plants across the country, connecting them with an energy grid for energy self-reliant villages, becoming a carbon–neutral society mitigating climate change and meeting sustainable development goals. Besides, India must develop a robust R&D network and facilities that connect all the stakeholders, such as industry, academic institutions, government, and people in this sector. It is necessary for skill development and capacity building in this domain.

Moreover, focusing on the indigenous development of low-cost biogas upgradation technologies is necessary, ensuring that their components and parts are locally available/manufactured and do not rely on imports from foreign countries. For example, IIT Delhi has also worked on developing a mobile biogas upgrade unit. This mobile unit can roam around in villages to biogas plants with surplus gas availability. It draws surplus biogas from the plant, upgrades it on its mobile unit, and compresses and fills it in cylinders on this unit itself. This can become a viable solution when many biogas plants work in nearby villages. This kind of unit can have a good entrepreneurship model where an entrepreneur can earn a good return on this gas to the family and then can use this upgraded biogas to run tractors, motorcycles, and other automobiles in rural areas running on CBG.

Furthermore, like the unconventional approach of using GIS in bolstering the CBG sector, as discussed earlier in the chapter, as a country, we must focus on developing alternate pathways to make CBG a lucrative business for investors and CBG entrepreneurs. One such application is green hydrogen production from CBG.

India aspires for "*Panchamrit*" to achieve net-zero emissions targets. It includes targets to achieve by 2030 as: 1) to have non-fossil fuel capacity of 500 GW, 2) to meet 50% of India's energy demand by renewable energy sources, 3) to reduce one billion tons of carbon emissions, 4) to reduce carbon intensity by less than 45%, and 5) to become carbon neutral and achieve net-zero emissions by 2070. India's journey towards net-zero emissions needs a targeted approach toward a green hydrogen ecosystem. Green hydrogen can fulfill rapidly increasing energy needs and decarbonize hard-to-abate sectors like refinery, steel, cement, fertilizers, and mobility towards the path to net-zero emissions. Indian government launched the "National Green Hydrogen Mission" (NGHM) on January 4, 2023. The primary aim of the Mission is to establish India as a Global Hub for the production, use, and export of green hydrogen and its derivatives, with a target of producing 5 MMT of green hydrogen year by 2030 [10, 11].

However, the most widely used fuel for hydrogen production is natural gas. The steam methane reforming method is mainly used for hydrogen production from natural gas and is associated with 10–12 kg CO_2-equivalent/kg H_2 emissions. The CO_2 emissions associated with hydrogen production and use has reached about 920 Mt CO_2 by 2023 [12–14].

As per the IRADe report, the electrolyzer capacity required to fulfill the National Green Hydrogen Mission target of 5 Mt of green hydrogen is between 60 and 100 GW. By 2050, to achieve a hydrogen demand target of 28.7 Mt, as outlined in the NITI/RMI report, an increase in electrolyzer capacity of 6–8 times (462–603 GW) is necessary, indicating significant market and investment potential for the electrolyzer installation and manufacturing sectors in the country. Hydrogen production will significantly increase electricity demand [15].

Excess CBG availability in many plants under the SATAT/GOBARdhan scheme of the GoI can pave an alternate pathway to green hydrogen production, bolstering the economic viability of a CBG plant. Besides, the end users of gray hydrogen, like chemical industries, steel industries, fertilizer industries, refineries, etc., can become the potential off-takers of CBG, apart from OMCs, switching to green hydrogen via CBG with the help of existing commercially available steam methane reforming technology, as a replacement for natural gas in hydrogen production. It can act as a "Universal solution" for all the underlying issues and targets.

For biogas upgrading technology, one thing that should also be taken care of is the loss of methane in the process, which should be minimized as per international protocols. The methane loss should not be more than 2%. Moreover, the stakeholders involved in the CBG sector should find alternate and sustainable solution to Carbon Dioxide Recovery and its application in greenhouse farming and grain storage warehouse, which is usually vented out in environment during the upgradation process.

In addition, to make CBG a success story, one must focus on utilizing CBG plant slurry as fermented organic manure (FOM) and liquid fermented organic manure (LFOM) as a sustainable approach to enhance soil health and additional viability of the CBG plant. Integrating FOM and LFOM into farming practices aligns with the principles of regenerative agriculture, focusing on soil regeneration, biodiversity enhancement, and ecosystem restoration. FOM and LFOM are key drivers of sustainable and regenerative agricultural systems, contributing to improved livelihoods, food security, and climate resilience.

The rising demand for compressed biogas (CBG) in India has created an opportunity to increase the production of organic digestates or effluents, such as solid and liquid fractions known as fermented organic manure (FOM) and liquid fermented organic manure (LFOM), respectively. These organic digestates, which remain after anaerobic biogas production, are rich in organic carbon and essential nutrients. Their application can reduce the reliance on chemical fertilizers, improve soil health, and mitigate the adverse effects of synthetic inputs [16].

The quality of FOM and LFOM depends on the type of organic feedstocks used, which influence nutrient composition, physical properties, total solids, volatile solids, ash content, lignin content, and the quantity of biogas produced during anaerobic fermentation in biodigesters [17, 18]. Another critical quality indicator is the fermentation period, with an optimal duration suggested to be around 40 days [19]. The effectiveness of FOM and LFOM can be further improved by adding different microbial consortia, seaweed extracts, plant growth regulators, biostimulants, humic acid, organic minerals, and leguminous crops. Such modifications can further enhance their effectiveness in promoting sustainable agricultural practices.

FOM and LFOM are produced through anaerobic digestion in oxygen-free biodigesters. In this process, microorganisms break down organic matter without oxygen, generating biogas and nutrient-rich digestate. The digestate is then separated into solid and liquid fractions using a solid–liquid separator. The liquid fraction (LFOM) is collected in large storage tanks or lagoons, and the solid fraction (FOM) is transferred to a platform, formed into heaps, and later arranged into windrows for final processing. The solid digestate is regularly turned and mixed to ensure proper decomposition to maintain adequate aeration. Temperature is monitored within the optimal 50–65 °C range to facilitate microbial activity. Different feedstocks yield FOM and LFOM with distinct physical and chemical properties. For instance, paddy straw-based FOM is rich in silicon, which benefits crop health.

As of September 2024, the Central Government has authorized 21 manufacturers to sell FOM and LFOM in bulk directly to farmers. This authorization, valid for three years, aims to promote the use of organic fertilizers and reduce dependence on chemical fertilizers. These guidelines underscore the Government's commitment to promoting sustainable agricultural practices through organic fertilizers, ensuring quality control, and providing financial incentives to encourage their adoption.

The Ministry of Agriculture and Farmers Welfare, through an order dated February 17, 2025, has made a special provision stating that organic material produced through the fermentation process in Compressed Biogas (CBG) plants—whether as a main product or a by-product—shall be considered an organic carbon enhancer and will be included under Schedule VIII of FCO. A summary of compliance and marketing guidelines is illustrated in Table 10.7.

Moreover, a marketing strategy for Fermented Organic Manure/Liquid Fermented Organic Manure must be developed to target farmers' needs, create value addition

Table 10.7 Summary of compliance and marketing guidelines

S. No.	Area	Requirement
1	Registration	Mandatory on Unified GOBARdhan Portal
2	Quality assurance	Testing in Govt./NABL-accredited labs
3	Packaging	Allowed in both packaged and bulk/loose forms
4	Marketing	Through fertilizer companies or direct-to-farmers
5	Eligibility for MDA	Only for products meeting FCO standards and under the GOBARdhan scheme
6	Duration of bulk sales pilot	Initial three years, extended till May 2026

in products, focus on R&D, capacity building, and increase awareness, branding and packaging, and proper selection of distribution channels.

To summarize, a robust ecosystem for CBG plants should be installed where raw material supply is not an issue and finished products like methane, carbon dioxide, and biogas plant spent slurry (FOM/LFOM) are adequately utilized in the nearby areas. This will pave the way for *Swachh Bharat Abhiyan to manage waste and create a win–win situation properly.*

10.8 Conclusion

India must prioritize compressed biogas (CBG) to preserve the environment and provide sustainable energy. CBG is a key answer as the country works to strike a balance between its commitment to lowering greenhouse gas emissions and its expanding energy demands. It provides prospects for rural economic development while controlling organic waste and generating clean, renewable electricity.

Initiatives by the government, such as financial incentives, budgetary allocations, and policies from several ministries, offer a solid basis for expanding the CBG industry. These steps are essential for overcoming logistical problems, regulatory complexity, and technical obstacles. For CBG projects to be successful, sites must be carefully chosen, considering infrastructure and feedstock availability.

India is steadily progressing toward energy self-reliance by promoting innovation and private-sector involvement. The country's objectives of reaching net-zero emissions and transitioning to a circular economy are supported by the inclusion of CBG in the energy mix. With persistent work, CBG can significantly contribute to India becoming ecologically sustainable and energy secure, aligning with the goal of "Aatmnirbhar Bharat" and a developed country by 2047. This route is an outstanding example of how ecological stewardship and economic advancement can work in synergy.

References

1. International energy agency (2021) India energy outlook 2021 world energy outlook special report. www.iea.org/t&c/
2. Jain R (2023) Greening India's energy mix with compressed biogas (CBG). https://www.cseindia.org/greening-india-s-energy-mix-with-compressed-biogas-cbg--11705
3. Ministry of Jal Shakti (2024) GOBARdhan : home. https://Gobardhan.Sbm.Gov.In/
4. IOCL (2021) Satat : home. https://Satat.Co.in/Satat/#/
5. Ministry of new and renewable energy (2024) MNRE : bio energy - schemes - waste to energy. https://Mnre.Gov.in/En/Waste-to-Energy/
6. Cabinet committee on economic affairs (CCEA) (2023) Unique package for farmers. https://www.Pib.Gov.in/PressReleasePage.Aspx?PRID=1935893
7. Ministry of petroleum and natural gas (2024a) BAM : home. https://Bam.Eil.Co.in/Default
8. Ministry of petroleum and natural gas (2024b) DPI : home–development of pipeline infrastructure for injection of CBG. https://Dpi.Eil.Co.in/Default
9. Jain R, Yadav NK (2024) Developing an effective state-level compressed biogas policy in India developing an effective state-level compressed biogas policy in India current status and key strategies policy brief. https://www.cseindia.org/developing-an-effective-sate-level-compressed-biogas-policy-in-india-current-status-and-key-strategies-policy-brief-12380
10. PIB (2023d) Press release: Press Information Bureau
11. Braj NS, Ratn T, Jha P (2024) India's green hydrogen revolution-an ambitious approach
12. Anon B (n.d.) Hydrogen resources|department of energy. Retrieved 13 April 2025. https://www.energy.gov/eere/fuelcells/hydrogen-resources
13. DST (2020) India country status report on hydrogen and fuel cells
14. IEA (2024a) Global hydrogen review 2024—Analysis–IEA
15. IRADe (2024c) Pathways for the green hydrogen development in India
16. Sogn TA, Riley H, Abrahamsen G, Krogstad T (2018) Effects of digestate and compost on soil fertility and plant growth. Agr Ecosyst Environ 256:204–210
17. Ndubuaku UM, Akinrinlola RJ, Ojo TO (2013) Physicochemical properties of organic wastes for biogas production. Int J Agric Res 8(2):74–81
18. Walsh JJ, Jones DL, Edwards-Jones G, Williams AP (2012) Replacing inorganic fertilizers with biosolids and anaerobic digestates: impact on soil properties. J Environ Manage 95(1):155–162
19. de Oliveira Neto CH, Figueiredo CC, Carneiro JS (2017) Optimal fermentation period for organic fertilizers derived from biogas digestates. Soil Plant Nutr J 30(3):420–432
20. BIS. (2025, August). *Indian Standard Biogas (Biomethane)-Specification (Second Revision)*. Bureau of Indian Standards. https://www.bis.gov.in

MIX
Papier aus verantwortungsvollen Quellen
Paper from responsible sources
FSC® C105338

If you have any concerns about our products,
you can contact us on
ProductSafety@springernature.com

In case Publisher is established outside the EU,
the EU authorized representative is:
Springer Nature Customer Service Center GmbH
Europaplatz 3, 69115 Heidelberg, Germany

Printed by Libri Plureos GmbH
in Hamburg, Germany